3

621.381 TUR

Microelectronics NII

David Turner BSc(Eng) CEng MIEE ACGI

Dean of the Faculty of Technology
College of Further Education, Plymouth

Stanley Thornes (Publishers) Ltd

First published in 1991 by:
Stanley Thornes (Publishers) Ltd
Ellenborough House
Wellington Street
CHELTENHAM GL50 1YD
United Kingdom

Reprinted 1993
Reprinted 1994

British Library Cataloguing in Publication Data
Turner, David
    Microelectronics NII. – (Microelectronics)
    I. Title   II. Series
    004.16

    ISBN 0-7487-1119-8

Typeset by Florencetype Ltd, Kewstoke, Avon
Printed and bound in Great Britain at The Bath Press, Avon.

Contents

Foreword

In 1982 Hutchinson Educational Publishers, now part of Stanley Thornes, published on behalf of the Business Technician Education Council (BTEC), a series of books designed for use as learning packages in association with the published standard units in Microelectronic Systems and Microprocessor-based Systems.

The last decade has seen the transformation of industrial computing with the explosion in personal computing. The reduction in price complemented by a significant increase in computing power has extended the application of personal computers so that microelectronics is now a realistic tool in all sectors of industry and commerce. The need for adequate training programmes for technicians and engineers has increased and the BTEC units have been revised and updated to reflect today's needs.

Stanley Thornes have produced a series of learning packages to support the updated syllabuses and numerous other courses which include microelectronics and Microprocessor-based Systems. There are five books in the series:

Microelectronic Systems	Level F	by G. Cornell
Microelectronics NII	Level N	by D. Turner
Microelectronics NIII	Level N	by D. Turner
Microprocessor-Based Systems	Level H	by R. Seals
Microcomputer Systems	Level H	by R. Seals

Two additional books which complement the five above are:

Microprocessor Interfacing	Level N	by G. Dixey
Practical Exercises in Microelectronics	Level N	by D. Turner

This book follows the BTEC unit and introduces the student to microprocessor systems. A second book *Microelectronics NIII* builds on the foundation of this text to give a complete coverage of the two N level units for BTEC students. Further books in the series by R. Seals extend the subject matter to more advanced system studies. *Practical Exercises in Microelectronics* complements this and the level NIII books.

Andy Thomas
Series Editor

Preface

This book is written to cover the objectives of the BTEC Unit Microelectronic Systems U86/333 which are specified for a single unit at NII level. The companion volume in the series covers the remaining objectives of the same unit at NIII level. In addition the two books are supported by a third, entitled *Practical Exercises in Microelectronics* which provides a large number of relevant practical exercises designed specifically for student use. The three Microelectronics books cover all aspects of the BTEC Unit from a theoretical and practical point of view providing material for both classroom activities and self-supported study.

For those students who prefer to study by distance- or open-learning methods, two Learning Packages are available which integrate the theory and the appropriate practical activites. The Packages also contain all the required hardware and software needed for a complete microelectronics course. Details are available from:

Plymouth Open Learning Systems Unit,
College of Further Education,
Kings Road,
Devonport,
Plymouth PL1 5QG
Telephone: (0752) 551947

The aims of this book are to introduce the student to microprocessor systems, their component parts and their interrelated functions. An introduction is also provided into the methods of microprocessor programming in machine code and assembly language.

Throughout this book and the others in the series, the majority of the examples and all of the programs use the Z80 microprocessor. Not only is the Z80 the most widely used 8-bit microprocessor, but it also illustrates most of the aspects of microprocessor operation which are in widespread use throughout industry. Its features such as the large number of registers, many addressing modes, non-multiplexed buses and the range of interrupt facilities permit detailed examination of these microprocessor features with reference to a single device. The large number of Z80-based computer systems in educational establishments can also be employed to run the programs contained in this and the other books in the series.

David Turner
March 1991

Acknowledgements

The author wishes to thank his wife Kathy, whose help and encouragement during the production of this book have been a constant source of inspiration and whose sense of humour has somehow managed to remain intact throughout. Thanks also go to his understanding children Beth, Alex and Ben, who should have seen a lot more of their father than they have done.

His thanks go to Jackie Boyce who typed and edited the manuscript so efficiently and whose typing speed is a good test for any wordprocessor. Thanks also go to Roger Bond and Elizabeth Frederick-Preece of POLSU who created some excellent illustrations from his very rough sketches.

The microprocessor based system

1.1 INTRODUCTION TO SYSTEMS

The title of the BTEC Unit which this material is based on is 'Microelectronic Systems', so it would be appropriate first of all to consider what is meant by a **system**. A simple dictionary definition of a system is: 'a collection of parts which work together as a collected whole'. This definition embraces many types of system, and the following are a few examples to illustrate the breadth of the definition.

- *Hi-fi system* – A system to reproduce music.
- *A washing machine* – A system to clean clothes.
- *The education system* – A system to educate children and adults.
- *The electric power system* – A system which provides electricity to industry and homes.
- *A motor car* – A system to provide transportation.
- *A microwave oven* – A system for cooking food.

There are many types of system and each may be very small or very large. Often the larger systems can be broken down into a number of smaller systems, which are referred to as sub-systems.

For example, a Hi-Fi system may be broken down into the following sub-systems:

- Tape deck.
- Disc player.
- Graphic equaliser.
- Tuner.
- Power amplifier.

Each of these sub-systems is actually a system in its own right and may also be sub-divided into further sub-systems.

One of the main features of any system is that it has **inputs** and it has **outputs**. It **processes** the inputs to produce the outputs.

These inputs and outputs may take one or more forms, for example they may be:

- Physical things.
- Energy.
- Information.

A very general representation of any system is shown in *Figure 1.1*. This shows how the system inputs are processed to produce the system outputs. For most systems the outputs depend upon the inputs and the process that takes place in the system. This process is fixed and is generally built into the system.

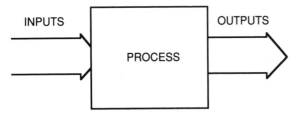

Figure 1.1   A general system block diagram

Consider, for example, a simple immersion heater as shown in *Figure 1.2*. The immersion heater is a device that processes cold water and electricity and produces hot water as a result. In common with all electric systems, the immersion heater must have an input of electric power in order to function. This is also true of other electric systems such as the Hi-fi system, a television, a computer, or a washing machine.

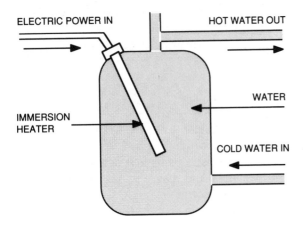

Figure 1.2   Simple immersion heater system

1.2 MICROELECTRONIC SYSTEMS

Before the advent of the computer all electric systems were the same as any other system, in that their function was fixed and determined by the design of the system. It could not be changed

unless it underwent a complete re-design of the circuitry or the component parts.

Some of the earliest types of TV games fell into this category. They were simple bat and ball games where a spot of light on the TV screen, the 'ball' moved from side to side and bounced off the 'bats' controlled by the players with two control knobs. This simple TV tennis arrangement was a design that used integrated circuits and was completely dependent upon the design of the electronic circuit for its operation. Extra features could only be added to the game if extra components were added to the circuit. For example, an extra switch would allow changes in speed in the game and another switch would allow changes in the size of the bat or the points scored.

This is in stark contrast to some of the modern video games, or computer games. In a modern video game the basic game circuit electronics remains the same but the complete function of the game can be changed by inserting a different program into the machine. Effectively this changes the process that is carried out by the machine although the inputs and the television output are similar. The reason is that a modern video game is based upon a microprocessor system in which the function or process carried out depends not only upon the design of the system but also upon the program that is stored in the microprocessor memory – or in the case of a game, in a cartridge.

Thus in microelectronic systems the **process** depends upon the **microprocessor program**. This is illustrated in *Figure 1.3*.

The program that determines the operation of a microprocessor based system is simply a set of coded instructions for the microprocessor. It is generally resident within the system, although it may be readily changed either by cartridge or by some other means. Therefore the microprocessor based system can perform many different tasks depending upon the program that it is operating. This concept of the same hardware being able to perform many different tasks is fundamental to the rise in the importance of microelectronics throughout all industry. It is also the reason why the microprocessor is so cheap. This arises because the same microprocessor can be put to work in thousands of different applications simply by changing the program which it is given to

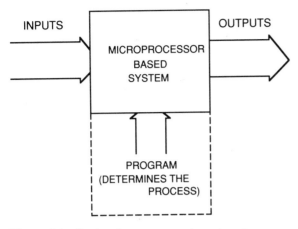

Figure 1.3 Basic microprocessor based system

operate. Because the microprocessor itself does not have to be modified in any way for different tasks, it can be mass produced and thus be made very cheaply.

Consider, for example, an electronics system that is used to control a set of traffic lights. Before the advent of microprocessors every set of traffic lights was controlled by a series of relays and these had to be individually connected into systems which were tailored for specific locations. Individual timing elements had to be calculated and incorporated into each circuit.

Now under microprocessor control, the hardware for every location is the same so that the microprocessor directly switches lights on and off but all the sequencing and timing operations can be carried out under computer control. Thus to modify the system for any location is simply a matter of changing the program. The hardware, including the input circuits and the light driving circuits, remains the same.

1.3 ELECTRONIC SIGNALS

Electronic circuits and the signals that are used within them can be broadly classified into two groups, **analogue** and **digital**.

Most physical quantities vary continuously which means that they are capable of infinitely small changes. When they are translated into electronic signals these also vary continuously and are known as **analogue quantities**. Typical examples include weight, temperature, length,

speed etc. Within an electronic circuit, these quantities would normally be represented by a voltage or current which varied between maximum and minimum limits, and could take on any value between the set limits. Conventional electronic circuits – such as amplifiers, filters, attenuators, etc. – are used to manipulate analogue quantities.

Digital electronic circuits are those in which the quantities represented can only have fixed values. Generally the term **digital electronics** refers to the types of circuit in which numbers and quantities are represented in binary form and therefore the circuit voltages and currents are limited to one of two values. Within a digital system, since signals may only have one of two values, they are generally represented by the numbers '0' and '1'. This allows digital signals to be manipulated using **binary mathematics**. *Figure 1.4* shows the difference between an analogue signal and a digital signal. It shows how the analogue signal can take on any value between its maximum and minimum value but the digital signal can adopt only one of two possible states.

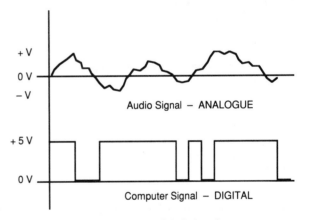

Figure 1.4 Analogue and digital signals

Computers can handle only digital signals, so that when analogue quantities need to be represented in a computer they first have to be converted to digital format. In addition the amplitude of the signals within a computer is strictly limited. Generally the signals must be between 0 and +5 volts, as shown in *Figure 1.4*. Fortunately, the process of converting between analogue and digital signals is relatively straightforward, which

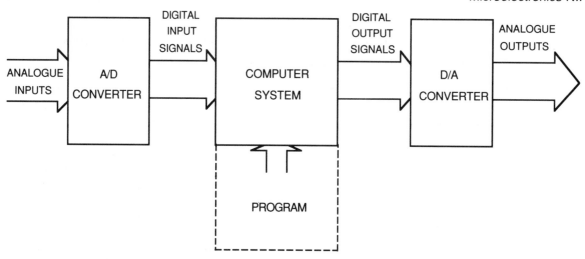

Figure 1.5 Analogue input and output to a computer

allows computers to handle analogue information in digital format and then produce analogue outputs. The device that converts analogue to digital signals is known as an analogue to digital converter, and the device that performs the opposite process is known as a digital to analogue converter. *Figure 1.5* shows how these two devices may be connected to a computer system so that it can input and process analogue signals and produce an analogue output, yet still maintain all the advantages of digital signal manipulation within the computer system.

1.4 BINARY AND HEXADECIMAL NUMBERS

Within a computer system all of the signals have one of two states, and it is convenient to represent these states as a Logic '1' (about +5 volts) and a Logic '0' (about 0 volts). The Logic '1' and Logic '0' are referred to as binary digits or bits for short. Thus a bit can be either '0' or '1'. If these were the only two numbers that could be represented by a computer then digital electronics would be severely limited. However, by combining binary digits in groups much larger numbers can be represented and this is the basis of all digital computing. *Table 1.1* shows the binary equivalent of a range of numbers from 0 to 20. Generally, in

Table 1.1 Binary equivalents to decimal numbers 0 to 20

Decimal	Binary	Decimal	Binary
0	0 0 0 0 0	11	0 1 0 1 1
1	0 0 0 0 1	12	0 1 1 0 0
2	0 0 0 1 0	13	0 1 1 0 1
3	0 0 0 1 1	14	0 1 1 1 0
4	0 0 1 0 0	15	0 1 1 1 1
5	0 0 1 0 1	16	1 0 0 0 0
6	0 0 1 1 0	17	1 0 0 0 1
7	0 0 1 1 1	18	1 0 0 1 0
8	0 1 0 0 0	19	1 0 0 1 1
9	0 1 0 0 1	20	1 0 1 0 0
10	0 1 0 1 0		

digital computers groups of bits of either 8, 16 or 32 are used.

For the numbers 0–15 the first binary digit in the number is always 0. The other four digits can be combined in 16 different ways, so that the numbers 0–15 can be represented by 4 bits. Similarly, when five bits are used the numbers up to 31 can be represented uniquely. *Table 1.2* shows the relationship between the number of bits used and the decimal number that can be represented by those bits.

In simple computer systems the numbers are generally represented using 8 bits. *Table 1.2* shows that this would allow 256 different numbers to be used. However, within a computer system it

Table 1.2 Relationship between number of bits and the decimal numbers represented by them

Number of bits	Decimal numbers represented	Number of bits	Decimal numbers represented
1	2	9	512
2	4	10	1024
3	8	11	2048
4	16	12	4096
5	32	13	8192
6	64	14	16 384
7	128	15	32 768
8	256	16	65 536

is also necessary to represent other data such as letters or computer instructions. Therefore special binary codes are used to represent these different items.

A special code known as the American Standard Code for Information Interchange (ASCII) is used to represent alphanumeric characters, i.e. all the letters, punctuation marks and figures normally found on a typewriter, as well as some special control codes. This ASCII code representation of alphanumeric information can easily be manipulated by a computer to allow messages to be transferred from one machine to another or to allow text to be manipulated within a computer system.

The computer itself requires certain instructions so that it can be made to operate in a particular way. These instructions are also in coded form and are stored in the computer memory. Codes are again groups of binary numbers, generally eight bits long, which are interpreted by the computer and taken to indicate a specific instruction. Groups of these numbers then form the computer program. Computer programs that consist of binary instructions are known as **machine code**. However computer programs are written, they always end up as a series of **machine code** instructions for the microprocessor.

Since 8 binary bits are required to represent computer instructions, a typical example of a piece of computer program may be illustrated below:

```
1 1 0 1 1 0 1 1
1 0 0 0 0 0 0 0
1 1 0 1 0 0 1 1
1 0 0 0 0 0 0 1
0 0 0 0 0 0 0 0
0 0 0 1 1 0 0 0
```

Clearly this type of code is very difficult to understand, interpret or remember. Therefore, a way of representing binary numbers in a simpler form is generally used in microelectronics. This is the method known as **hexadecimal** numbering.

Table 1.3 Hexadecimal number system

Binary	Hexadecimal	Binary	Hexadecimal
0 0 0 0	0	1 0 0 0	8
0 0 0 1	1	1 0 0 1	9
0 0 1 0	2	1 0 1 0	A
0 0 1 1	3	1 0 1 1	B
0 1 0 0	4	1 1 0 0	C
0 1 0 1	5	1 1 0 1	D
0 1 1 0	6	1 1 1 0	E
0 1 1 1	7	1 1 1 1	F

In the **hexadecimal** system, the binary numbers from 0 to 15 are given unique codes as shown in *Table 1.3*. Therefore, when binary numbers need to be represented, the bits of each number can simply be grouped in 4s and a hexadecimal number can be used instead.

For example,

1 0 1 1 0 0 1 1

can be written as

B 3

and

1 0 0 0 1 0 1 0

can be written as

8 A

Anyone who works in microelectronics must be able to understand the hexadecimal number system. The numbers from 0 to 9 are the same as the decimal numbers represented in binary, so the only difficulty is in remembering the letters. The easiest way is to remember:

A = 1 0 1 0

C = 1 1 0 0

F = 1 1 1 1

The other letters are then one more or one less than each of these.

Frequently it is necessary to be able to convert freely between **binary** and **hex**. The following examples illustrate the technique for numbers with more than 8 bits.

EXAMPLE 1
Convert the number 1 0 1 1 1 1 1 1 0 0 1 1 0 1 0 1 to hexadecimal.

Split the number into groups of 4 bits starting on the right:

1 0 1 1   1 1 1 1   0 0 1 1   0 1 0 1
  B   F   3   5

so 1 0 1 1 1 1 1 1 0 0 1 1 0 1 0 1 = BF35 in hexadecimal.

EXAMPLE 2
Convert 1 8 0 D to binary.

Write down the binary code for each hexadecimal character.

1   8   0   D

0 0 0 1   1 0 0 0   0 0 0 0   1 1 0 1

Thus 1 8 0 D = 0 0 0 1 1 0 0 0 0 0 0 0 1 1 0 1 in binary.

The instructions for a computer are always held within the system in the **computer memory**. It would be possible to pass the information from the memory to the microprocessor in groups of eight bits along a single wire. This would be known as **serial data transfer**. However, this would be far too slow for modern computer operations. Instead there are generally eight paths, one for each bit, between the computer memory and the microprocessor. This means that data can be passed in **parallel** format, which means that the computer can operate far more quickly. This idea is illustrated in *Figure 1.6*.

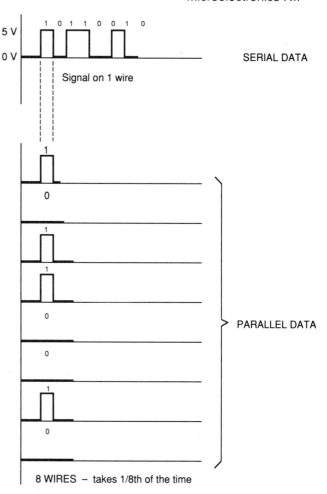

Figure 1.6   Serial and parallel data

1.5 MICROPROCESSOR BASED SYSTEM

Figure 1.3 (page 3) showed a basic microprocessor based system as a single block with inputs and outputs. The function of the system was determined by the program. The detail of what is inside the box can now be examined, and is shown in *Figure 1.7*.

Inside a microcomputer system, there are four basic elements. The inputs from the outside world enter the system via an input unit known as an **input port**. Similarly, the outputs which are connected to devices outside the computer are derived from an **output port**. The two other units within the microcomputer are the central processing unit (CPU) and the **memory** or store.

Generally, each of the blocks is a separate integrated circuit, although the **input port** and the **output port** may be combined in a single chip. Also there may be more than one memory chip. Some modern control computers, integrate all of the elements shown into a single integrated circuit which is known as a single chip microcomputer. However, most general-purpose computers have separate circuits for the elements shown above.

The four elements are interconnected via three buses. These are simply wire connections between all the parts of the circuit, and each bus has a separate function.

The function of each part of the circuit is as follows.

Input Port

The **input port** receives signals from external devices, known as peripherals. For example, a keyboard may send signals to the computer. These signals must always be in binary form, and must also be the low-level signals, between 0 and +5 volts, that the computer can deal with. In any computer there may be several input ports although only one is shown in *Figure 1.7* for simplicity.

Output Port

The **output port** sends binary signals, between 0 and +5 volts to external peripherals. For example, a computer output may be displayed on a screen, and the signals from the computer would need to be sent from an output port. Similarly, when a computer wishes to print information this must also come from an output port to a printer.

Central Processing Unit (CPU)

The central processing unit is the heart of any computer system. It is a device that performs all the calculations, makes the decisions, and controls the rest of the system. Although it is not shown in the diagram, the rate at which the CPU works is controlled via a 'clock'. This clock is generally a highly stable squarewave signal derived from a crystal oscillator. The frequency of this squarewave signal is generally between 1 MHz and 4 MHz in older types of computer. In more modern machines the clock frequency can be as high as 30 MHz. The CPU is the device that controls the three buses, the **address**, **data** and **control** buses.

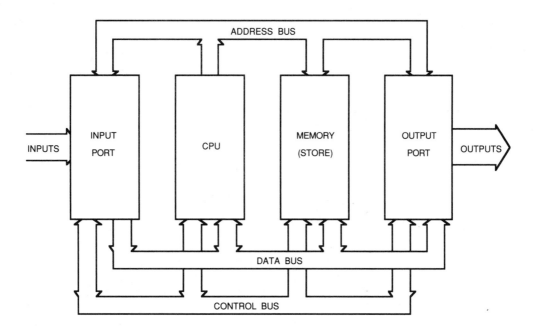

Figure 1.7   Microcomputer block diagram

Memory

The **memory** is the part of the system that stores both the instructions for the CPU and any data that it needs to work on. It is best to think of the memory as a series of separate store locations or pigeon-holes, each capable of storing one piece of information. The unit of information stored in a memory is generally measured in bytes and each memory location stores one byte. In all computers there are thousands of unique memory locations, and each one must be capable of being accessed uniquely. For this reason memory locations are each given an **address**. The **address** is the unique number of each memory location within the whole memory. Typically in small computer systems the maximum number of memory locations is 65 536. In more modern systems this number may be in excess of one million and, perhaps, as much as sixteen million bytes. An important thing to note about a computer memory is that it generally stores its data and instructions sequentially in addresses in ascending order. The job of the CPU therefore is generally to READ instructions from sequential memory addresses and perform the actions that are required by the program stored there.

The Bus System

Each of the blocks in the block diagram of the microprocessor system communicates with the other elements in the system by means of connections known as the buses. There are three buses in each system, known as the address bus, the data bus and the control bus. Each one has a unique function and together they ensure that the system functions correctly.

Address Bus

The **address bus** originates in the central processing unit. It is generally a group of sixteen connections which are used to hold the unique binary number during each cycle of the CPU operation. This binary number is the **address** of the part of the system or the memory location which is to be used during the CPU cycle. It is the address bus that tells the system which part is to be used at any moment. The address bus is derived from the CPU and goes to every other part of the system. It is therefore **unidirectional** in operation.

Data Bus

The **data bus** is used to carry information from the part addressed by the address bus back to the CPU, or from the CPU to any other part of the system. One end of any data transfer in small systems is always the CPU. Typically the data bus may carry information from the memory to the CPU or from the CPU to an output port. Because of this, signals may be transferred in either direction along the data bus although it is only used in one direction at any instant. However, the data bus is known as a **bidirectional** bus. In small systems data is transferred eight bits, or one byte at a time, and it is the size of the data bus that generally determines the size of the system, 8, 16 or 32 bits wide. In the Z80 Based System the data bus is 8-bits wide.

Control Bus

The **control bus**, as the name suggests, controls the operation of all the other parts of the system. Most of the signals are derived from the central processing unit. There are four major control signals, **read**, **write**, **memory request**, and **input/output request**. If the address bus indicates which part of the system is to operate at any moment, the control bus can be thought of as indicating how it should operate. For example, when information is required from the memory, the address bus provides the address of the location from which the information is required, the control bus indicates a **memory request and read** operation must take place and the data is returned to the CPU using the data bus. Most signals are derived from the CPU but a few are inputs to the CPU. The bus can therefore be thought of as being two directional. Different CPUs have different numbers of control signals and each one generally has a unique control arrangement. Typically there may be between ten and twenty control signals.

1.6 THE FETCH/EXECUTE CYCLE

Microprocessors are driven by a crystal oscillator at very high speed and they continue the relentless cycle which involves fetching instructions from the memory, decoding them, and then executing each instruction as it is decoded. This relentless cycle is known as the fetch/execute cycle (*Figure 1.8*). Even when a microprocessor appears to be doing nothing, it is executing instructions at a very high speed. Most of the time, the instructions are stored in the computer memory in sequential order so that the fetch/execute cycle merely consists of fetching one

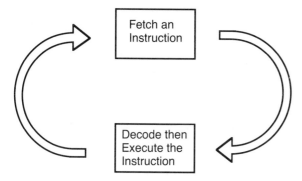

Figure 1.8   The fetch/execute cycle

instruction after another from the sequential addresses in the memory. However, from time to time an instruction is fetched which causes the processor to go to a different part of the memory to fetch the next group of instructions. A typical instruction will be executed by the CPU in three distinct phases, as follows:

(a) Fetch.
(b) Decode.
(c) Execute.

The Fetch Phase

At the beginning of each instruction the microprocessor will place a 16-bit code on the address bus, which will be the address of the memory location that contains the instruction. At the same time a memory **read** signal will be placed on the control bus so that the addressed memory location containing the instruction is read. The data from this

memory location is then placed on the data bus by the memory and will enter the CPU where it is to be stored.

The Decode Phase

When the data from the memory is stored in the CPU it is decoded to identify the instruction which the CPU is required to execute. Most CPUs can execute hundreds of different instructions and so this decode phase takes place in some internal logic circuits.

The Execute Phase

Once the instruction has been decoded the microprocessor will know exactly what sequence of events to initiate in order to execute the instruction. The execute phase may involve just an internal CPU operation, e.g. add two numbers together, or the instruction may require additional memory or input/output **read** or **write** phases. For example, the CPU may be required to output data to an output port, in which case the address of the output port would be placed on the address bus, an output **write** signal would be placed on the control bus, and the data to be written to the output port would be placed on the data bus.

It can be seen from the description above that the fetch and decode phases will be the same for any instruction, but the execution phase will vary in complexity and timing dependent upon the nature of the instruction required. Some instructions will take longer to execute than others, and consequently the length of time it takes to execute a program depends entirely on the complexity of the instructions contained in the program and the speed of the clock frequency to the CPU.

1.7 CPU INTERNAL OPERATION

Before considering the operation of the complete CPU, it is worth investigating the operation of one element of it that is used to store information. This is known as a **register**.

Register Operation

Microprocessors are built from the same basic logic elements contained in all digital circuits. In

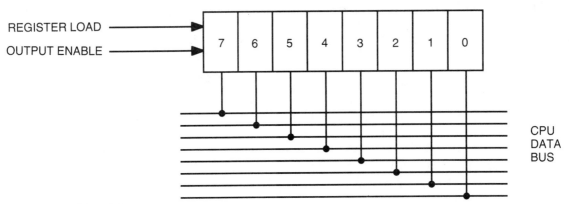

Figure 1.9   8-bit register

their simplest form, these include AND, OR and NOT gates. Using them as building blocks, it is possible to make more complex logic circuits such as **decoders**, **counters** and **registers**, which are themselves used in the construction of the microprocessor.

A **register** is a digital storage device capable of storing a group of bits. Integrated circuit registers normally store either 4 or 8 bits of data. However, the registers which form part of a microprocessor usually store either 8 or 16 bits of data. In many ways, a microprocessor register is very like a memory location except that it is normally designated by a name or letter rather than address.

Consider the register in *Figure 1.9* which is connected to the data bus inside the CPU. Each bit of the register is connected to the corresponding data bus line. There are two control signals, **register load** and **output enable**.

When the **register load** signal is active, data from the data bus is stored in the register. This is sometimes known as writing data into the register. When the register is loaded, any previous data is automatically overwritten.

The other control signal is the **output enable**. When this signal is active, data from the register is put on to the data bus so that it could be transferred to another part of the system. In the microprocessor both these control signals originate in the Control Unit.

Z80 registers

The Z80 microprocessor has a number of special registers such as the program counter which will

be dealt with later. It also has two sets of general-purpose registers known respectively as the main and alternate register set. There are six registers in each set although the vast majority of the microprocessor instructions operate only on the main registers (*Figure 1.10*).

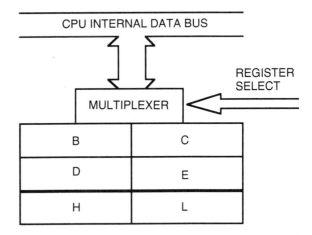

Figure 1.10   Z80 main registers

Each register is designated by a letter as far as the programmer is concerned, although internally, the microprocessor control unit uses a 3 bit address for each one. They are connected to the CPU internal data bus via a digital switch known as a **multiplexer**. This places the restriction on the system that only one register may be connected to the data bus at any time. Generally the accumulator is referred to as register A, and then registers B, C, D and E follow on alphabetically. There are historic reasons for calling the last two registers H and L since they stand for High and Low respectively. Previous microprocessors made by the same manufacturers used the H and L registers to hold the High and Low parts of a memory address when this was needed in an instruction. The Z80 maintained this feature and hence has kept the same register names.

The Intel 8080 and the 8085 microprocessors have an identical **main** set of registers, although they do not contain an alternate register set. This is because both the Z80 and the Intel 8085 were derived from the Intel 8080. By keeping the same main register set, it meant that any software written for the 8080 microprocessor, such as the CP/M operating system, would also run on the Z80 or Intel 8085.

The **accumulator** is a special register in the CPU. It is always used to store one of the inputs and the results of CPU calculations and, therefore, can be said to 'accumulate' the answers. Think in terms of a long calculation whose final total is gradually accumulated in this register. Naturally, not all calculations are very long, but that is how the name originated.

In *Figure 1.10* the registers have been deliberately drawn in pairs since there are a number of instructions which allow two 8-bit registers to be

combined as a 16-bit register pair. These are the BC, DE and HL pairs.

Whether used on their own, or in pairs, all of the general purpose registers are capable of storing data which may be required for calculations, etc. in the CPU. Data held in the registers may be accessed very much more quickly than data held in memory locations, although the restriction that only one register may be connected to the data bus at any time via the multiplexer still applies.

The Z80 CPU

Figure 1.11 shows the internal structure of the Z80 microprocessor. Some parts of the CPU have already been described but an explanation of their function has been included here for completeness.

Internal Bus Structure

The internal data bus is simply a continuation of the system Data Bus inside the CPU. It carries data between different parts of the microprocessor. Since the Z80 has only one internal data bus it is said to be a 'single bus' microprocessor. Some more complex processors have two or even three

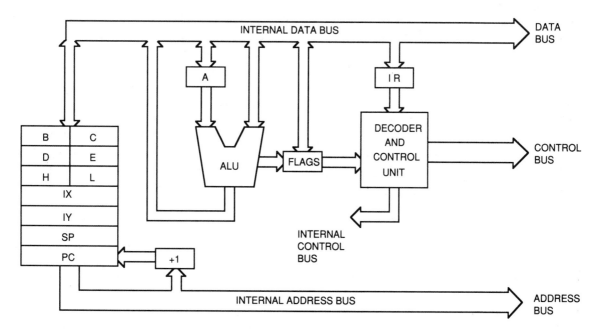

Figure 1.11   Z80 simplified internal architecture

internal data buses. This increases the processing speed but takes up more chip area.

The control bus is derived from the control unit inside the microprocessor. Although it is shown in the diagram with an arrowhead pointing in only one direction, a few of the control signals are actually inputs to the CPU. All the individual control bus lines are unidirectional. Note that there is also an internal control bus which controls all the CPU operations.

The address bus comes from the 16-bit address registers in the CPU and is connected directly to the external system address bus. Although the addresses generally come from the program counter, the contents of any 16-bit register or register pair may be placed on the address bus when required.

Instruction Register/Control Unit

The instruction register and control unit are the parts of the CPU that determine how it will operate. The control unit generates all the necessary internal and external signals to synchronise operations and itself is run by a 'microprogram'. Whenever an instruction is fetched from memory, it is clocked into the instruction register by the control unit. The instruction (8 bits) is decoded and then further control signals are set up to enable the instructions to be carried out. The decoding process translates each instruction into the correct sequence of control signals which must be sent out by the control unit to perform the required function.

The instruction may cause a purely **internal** change such as moving data between registers or performing some arithmetic operation on data in the registers. Alternatively it may cause an **external** operation, such as a **memory** read or write to take place. This is known as the **execute** phase of the operation.

Arithmetic Logic Unit/Accumulator

Whenever the instruction fetched requires an arithmetic or logical process to be carried out, the **execute** phase of the operation uses the ALU. In most microprocessors, the accumulator must contain one of the **operands** before the arithmetic or logical instruction is executed, and it is used to contain the **result** after the operation. In this context an operand refers to one of the data inputs to the arithmetic or logical process to be performed. The accumulator is also special in that there are a number of instructions which allow the accumulator to be used in ways that the other registers cannot be used. These include basic input/output operations and certain memory read/write operations.

The actual operations that the ALU can carry out are usually fairly limited. Typically these include:

(a) Add.
(b) Add with carry.
(c) Subtract.
(d) Subtract with carry.
(e) Increment.
(f) Decrement.
(g) Compare.
(h) Logical AND.
(i) Logical OR.
(j) Logical EXCLUSIVE OR.
(k) Rotate right.
(l) Rotate left.
(m) Complement.

Whenever the ALU is used, the result may either **set** or **reset** the **flags** in the flag register.

Flags

The results in the ALU which **set** or **reset** flags in the flag register, can then be examined by the control unit, and used as the basis of decision making if required. Most CPUs have five or six flags, each of which is simply a 1-bit store.

The Z80 flags are carry, half-carry, zero, sign, parity, overflow and add/subtract.

These are dealt with in more detail later.

Most arithmetic operations affect one or more of the flags in some way and the manufacturers data for the CPU should be consulted for details.

The control unit consults the flag register whenever a conditional instruction is read. It may be, for example, a jump on zero, which causes program execution to jump to a new address only if the zero flag is set.

General Purpose Registers (8 bit)

The six **main** general purpose registers are called B, C, D, E, H and L. They are shown in *Figure 1.10*.

They are used to store temporary working data to be used at some stage in arithmetic or logical operations. They are connected to the internal data bus via a multiplexer, so that only one has access to the bus at any time.

Some operations allow the registers to be used in pairs (16 bits) although the constraint of transferring 8 bits at a time over the data bus still applies. These pairs are BC, DE and HL. The HL pair has particular significance as an address pointer (H = Higher order byte, L = Low order byte) although they can be used for other purposes.

The HL pair is sometimes known as a memory address register.

Address Registers (16 bit)

In addition to the general purpose registers which may be combined to form 16-bit registers, there are also four 16-bit address registers. These are the program counter (PC), the stack pointer (SP), the X index register (IX) and the Y index register (IY). They all have special functions within the microprocessor.

Program counter (PC) This is one of the most important CPU registers since it always contains the CURRENT address from which the CPU has to receive instructions. Whenever an instruction is required from memory, the PC contents are placed on the address bus and the data returns from the memory on the data bus. Each time the PC is used it is automatically incremented and replaced in the PC register. This enables a CPU to systematically step through a program.

Whenever a jump instruction is encountered, the PC contents are replaced by the jump address, and program execution continues from the new address.

Stack pointer (SP) The STACK is generally a part of memory which can be used as a last-in, first-out storage device – like a stack of plates in a restaurant. Information to be stored on the **stack** is put into the memory location pointed to by the

stack **pointer**, which is automatically decremented before each byte of data is stored. Thus the stack pointer points to the last memory location used. If data is read off the stack, the stack pointer is incremented after each byte is read.

The stack pointer must be set up before any subroutines are used in a program.

Index registers (IX and IY) An index register is a general-purpose memory pointer which may be used for consulting data tables, etc. in memory. A data table is simply a list of binary numbers which may be required in a program. A general feature of an index register is that a 'displacement' can be added to it before its value is placed on the address bus. For example, if an index register points to the base of a data table in memory, the fifth item in the table may be consulted by adding 05 to the index register value before it is used. The actual value in the index register does not change as a result of the operation.

The two Z80 index registers both perform similar functions.

1.8 MEMORY

The memory in a computer system is vital for its correct operation. All the coded instructions that the microprocessor follows are held in the memory, generally in sequential locations. In addition, whenever an operation requires data, that data must also be held somewhere in the memory, so that although there is no physical difference between memory locations, some parts of it are used to store instructions and other parts of it are used to store program data. Instructions and data may be intermixed, and in fact some instructions include data in the memory addresses immediately after them. However, there are also occasions when a complete area of memory is reserved only for program data.

The maximum size that a memory can be is determined by the number of address lines that are connected to it. A memory size and the number of address lines are linked by the formula below:

$$\text{Memory size (bytes)} = 2^N$$

where N = a number of address lines.

So for example in a system which has 16 address lines, the maximum number of memory locations which can be accommodated in the system is:

$$\text{Memory size} = 2^{16}$$
$$= 65\ 536 \text{ bytes}$$

For modern microprocessors that have more than 16 address lines, clearly the number of bytes of memory that can be addressed will be considerably more. A system with 20 address lines could uniquely access over one million memory locations, while a system having 32 address lines can access over four thousand million memory locations.

The exact number of memory locations in a system is not usually important, so memory size is often given as 16K or 64K. This means about 16 000 or 64 000 respectively.

In fact, $1K = 1024$ in computer terms (2^{10}), so that:

$$2K = 2048$$
$$4K = 4096$$
$$8K = 8192$$
$$16K = 16\ 384$$
$$32K = 32\ 768$$
$$64K = 65\ 536$$

Each multiple is simply rounded down to the nearest 'power of two' in whole numbers of thousands.

Types of Memory

There are two main types of memory required in any microprocessor system. These are known as **read only memory** and **random access memory** respectively (*Figure 1.12*). Each type is

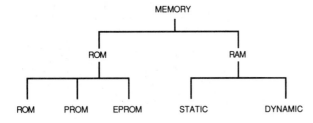

Figure 1.12   Types of memory

necessary for certain operations within the system, and all computer systems contain both types of memory though not necessarily in equal proportions. General-purpose computer systems contain far more **random access memory** than **read only memory**, whereas dedicated computer systems contain far more **read only memory**.

The major difference between the two types of memory is that **read only memory** does not lose its information when the power is removed, whereas **random access memory** loses everything it contains even if power is lost for a few microseconds. This means that at switch on, only the **read only memory** will contain useful information whereas the **random access memory** will contain 'garbage'.

The other important difference which follows from this, is that **read only memory** cannot have its data changed easily and some types of **read only memory** cannot be changed at all. However, the data within the **random access memory** can be changed very quickly and it is this fact that makes it possible to re-program computers with new information very quickly. Generally programs are loaded into **random access memory** from other storage devices such as magnetic tape or floppy disks. Only the parts of the computer program that never need to be changed are stored in the **read only memory**.

Read only Memory

All systems must contain some **read only memory** (ROM for short) so that they can perform some useful function as soon as the power is applied to a system. The ROM contains a program known as the **MONITOR** which stores the programs that enable the system to start correctly. For example, some of the programs within the monitor ROM will allow the microprocessor to read the keys that are pressed on the keyboard and decipher them, and to operate any display that is connected to the system. In addition they may contain programs to read and write information onto magnetic tape, sound the loudspeaker with a bleep, etc.

ROM   A true **read only memory**, as its name suggests is one which is programmed once, by the semiconductor manufacturer, and once this has taken place it is impossible to reprogram. It is the semiconductor manufacturer who puts the user's program within the device and this is achieved during the production process. Because of this and the economies relating to semiconductor manufacture it is only possible to produce ROMs in large quantities at an economical rate. Typically, to make ROM an economical proposition in any system, it must be anticipated that over 10 000 systems would be produced. This makes ROM attractive for some types of production process, for example in the manufacture of washing machines or other domestic appliances, but uneconomical as far as small-scale production is concerned. The main advantage of the ROM is that if large quantities can be produced then they are very cheap.

PROM PROM stands for **programmable read only memory**. This device is manufactured by a semiconductor manufacturer as a blank device, i.e. it contains no program, but it can be programmed by the end user. This allows the semiconductor manufacturer to maintain large quantities and therefore the devices are relatively cheap, while the end user is then able to program each PROM with his own program. This customises them for a particular application. This user programmability makes them attractive for relatively small quantities, say between 1000 and 10 000 items, since they are relatively more expensive than ROMs. The process of putting a program into a PROM is achieved in a special programming device known as a **prom blower**.

During the manufacture of the PROM the device contains a large number of small fusible links, these fuses can either be left intact or 'blown' during the programming process. They are blown simply by applying a high voltage pulse for a short time. Once blown the PROM cannot have its program altered and therefore any changes or errors made result in a useless device which must be discarded and replaced by the new or corrected program. Once a program has been finalised many manufacturers use PROMs to put into their equipment. However, since they are relatively expensive they are not particularly useful during the development phase of a product since every

time the program is changed they would have to be discarded.

EPROM An EPROM is an **eraseable programmable read only memory** (*Figure 1.13*). This means that although it can be programmed in a similar way to a PROM its contents can also be erased. The most common way of erasing an EPROM is to expose the silicon chip to shortwave ultra-violet radiation for about 15–20 minutes. To allow this to happen, the manufacturers leave a small glass window above the silicon chip in EPROM circuits, and they can therefore be readily identified. EPROMs often have their windows covered with a small piece of sticky paper. This is because the EPROM contents could be erased if

Figure 1.13 EPROM

the silicon chip is exposed to stray ultra-violet radiation for any length of time.

EPROMs are relatively expensive and would therefore not be generally used in large-scale manufacturing processes. However, for small-scale production, and for development work where the program may change frequently, the EPROM is an ideal way of storing programs.

Random Access Memory

Random Access Memory (RAM) has the property that all its information is lost as soon as the power is removed. However, because its contents can be changed readily then it has a vital function in all computer systems. The speed with which this can happen is known as the **memory access time**. As with the **read only memory**, there are a number of types of RAM although here, only two types will be considered, **static** and **dynamic**. These two types may often be used together in a system, although generally small systems tend to use static RAM whereas larger systems tend to use dynamic RAM.

Static RAM

Static RAM is based on a two-transistor memory element. In digital logic this memory element is known as a **bistable** and is arranged so that when one transistor in the bistable is conducting the other is non-conducting. Since the bistable can change state, i.e. the conducting transistor can change, it is capable of storing a binary 1 or binary 0. However, because two transistors are required for each memory element, the area of silicon taken up by each static memory cell is relatively large and this means that only relatively small memory sizes may be accommodated within a static memory chip.

In addition because one transistor is always conducting, static RAM tends to require a relatively large amount of power. This means that it gets hot, and also that relatively large power supplies would be required if a large amount of static RAM were incorporated in a system. It does have the advantage that it requires no additional circuitry other than the RAM chip itself to be included in a computer system.

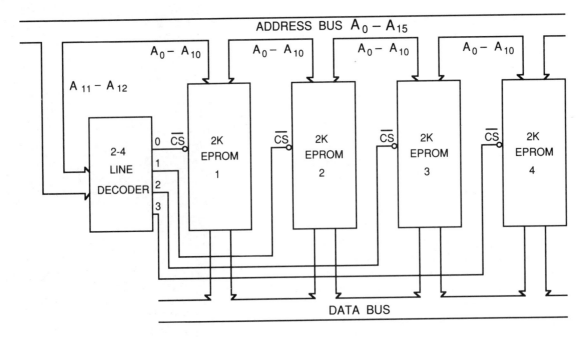

NOTE: $\overline{CS}$ Indicates Chip Select (logic 0 to select device)

Figure 1.14 Typical microprocessor system memory

Dynamic RAM

Dynamic RAM is based on a single transistor and storage capacitor memory cell. The storage capacitor stores a charge which can be either present or absent, thus presenting a logic 1 or a logic 0. The problem with this is that the charge on a capacitor leaks away relatively quickly, and the information therefore needs to be refreshed regularly. In practice the charge leaks away so quickly that the refreshing must take place every 2 ms. This is undertaken by additional circuitry used in conjunction with the dynamic RAM chips known as the **refresh circuitry** which must be present on any dynamic memory board used in a computer system.

Despite this limitation, dynamic memory is used extensively in large computer systems because the single transistor memory cell arrangement means that dynamic RAM chips can store a greater density of information than a static RAM chip. This means that the storage density is higher, the area of circuit board taken up is lower, and the cost is consequently lower for dynamic devices. In addition because the information is stored as the charge on a capacitor this requires relatively little current so dynamic memory devices require only a small power supply. Traditionally dynamic memory devices have been able to operate faster than static devices and the typical access time is of the order of 150 ns. However, modern developments mean that static devices now exist that are much faster than dynamic chips although they are relatively expensive.

Memory Arrays

Memory devices are made in a variety of different sizes, and it is unlikely that the complete memory for a microprocessor system can be implemented with a single memory chip. For example, a typical EPROM may store 2048 bytes of information out of a total of 64K bytes in the complete memory. Therefore to populate the memory with similar devices, 32 EPROMs would be required. Clearly these would all be manufactured identically so there must be some means of assigning each device a separate address within the complete microcomputer address space.

This function is achieved by the use of logic devices known as **decoders** and they allow independent addresses to be given uniquely to each memory chip in the system. By using the high order address lines as inputs to a decoder, each chip can be selected independently and therefore provide a complete coverage of all the addresses in the system. A simple memory system with four EPROMs and a decoder is shown in *Figure 1.14* (opposite).

Memory Maps

Figure 1.14 shows four EPROMs in a small microprocessor system memory. If they were the only memory devices in the system their addresses and the space they occupy in the complete microprocessor address space could be shown on a diagram known as a **memory map**. The memory map simply shows the 64K possible addresses for an 8-bit microprocessor system with blocks

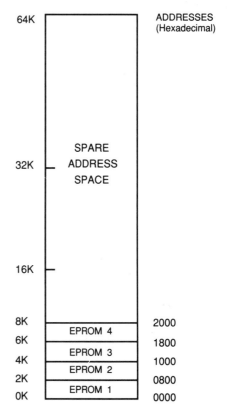

Figure 1.15 Simple microprocessor system memory map

occupied by those memory chips that exist. Alternatively, smaller versions of the diagram may show greater detail of the addresses occupied. *Figure 1.15* (page 17) shows how the memory devices in *Figure 1.14* could be represented on a simple memory map.

Generally, a microprocessor system will contain both EPROMs or PROMs and RAM devices, and there may be spaces in the memory map so that, for example, the RAM may start at a higher address than the top of the EPROM in the system.

Summary

The important points covered in this Chapter are:

- Microprocessor systems perform a function that is determined by their programming.

- Digital signals represented by 0 V and +5 V are used within a computer system.

- Binary numbers can be used to represent the digital signals, 0 V being a binary 0 and +5 V being a binary 1.

- Hexadecimal numbers are a convenient way to simplify the use of binary numbers.

- ASCII (American Standard Code for Information Interchange) Code is used to represent text in computer systems.

- All computers work on a **fetch, decode, execute** cycle.

- Registers are used to store data and addresses within the microprocessor.

- The main types of register in the Z80 micro-A processor are the Accumulator – A, General Purpose Registers – B, C, D, E, H and L, the FLAGS, Address Registers – Program Counter, Stack Pointer, Index Registers.

- The **program counter** is used to step sequentially through a program.

- The main types of memory are ROM and RAM.

- ROM devices hold their data even though power is removed, whereas RAM devices lose all their data.

- The data in RAM can be changed whereas data in ROM cannot.

- A **memory map** shows which type of memory is present at each address in a system.

Questions

1.1 Consider the following systems, and for each one write down the inputs, the process which is carried out, and the outputs:

(a) The telephone system.

(b) A radio receiver.

(c) A domestic toaster.

1.2 For the following list of devices, indicate whether the signal they produce is analogue or digital:

(a) The stylus of a conventional record-player.

(b) A doorbell.

(c) A thermostat.

(d) A pressure sensor.

1.3 What are the normal signal levels that represent the possible states in a digital computer?

1.4 Briefly describe the function of an analogue to digital converter.

1.5 List the three things that the binary numbers can be used to represent in a computer system.

1.6 Make the following conversions:

(a) 1 1 0 0 0 1 0 1 to hexadecimal.

(b) 1 1 1 1 0 0 0 1 to hexadecimal.

(c) 1C to binary.

(d) AF to binary.

1.7 Briefly describe a difference between serial and parallel transmission of data.

1.8 What is the purpose of ASCII code?

1.9 Briefly describe the function of the address bus in a microprocessor system.

1.10 Which of the buses in a microprocessor system is described as bidirectional, and why?

1.11 Briefly explain why some instructions take longer than others to execute.

1.12 What is the main function of a register?

1.13 Why does a register require two control signals to make it operate?

1.14 Why are the registers in a microprocessor connected to the data bus via a multiplexer?

1.15 What causes the **flags** in the **flag register** to change state?

1.16 What happens to the results of calculations performed in the ALU?

1.17 How does a microprocessor automatically step through a computer program?

1.18 What is the function of the block of general purpose registers?

1.19 If a system had 24 address lines calculate how much memory could be directly addressed by the microprocessor.

1.20 A manufacturer has to produce five specialist instruments for a customer. Which would be the best type of read only memory device to use in the system?

1.21 List the advantages that static RAM has over dynamic RAM.

1.22 Briefly describe the function of the program counter register in a Z80 microprocessor.

1.23 Convert the following binary numbers into hexadecimal:

(a) 0 0 0 1 1 0 0 0 1 1 1 1 0 0 0 1

(b) 1 0 0 0 0 1 0 1 1 0 1 0 1 1 0 0

(c) 0 1 1 1 1 1 1 1 0 0 1 1 0 1 1 0

1.24 Convert the following hexadecimal numbers to binary:

(a) 1989.

(b) 56AD.

(c) FAB4.

1.25 Explain why it is necessary to have ROM in a microprocessor system.

TWO

The fetch/execute sequence

OBJECTIVES

*When you have completed this chapter, you should
be able to:*

1. *Explain the basic operation as fetching the instruction to the microprocessor, decoding the instruction within the microprocessor, fetching more data if required and executing the instruction.*

2. *Illustrate the fetch–execute sequence for a simple data transfer instruction involving the accumulator and memory or input/output port.*

3. *Illustrate the execute sequence for a simple jump instruction.*

4. *Interpret timing diagrams to show the relationship between clock pulses and bus signals for the data transfer in (2).*

2.1 STORED PROGRAM CONCEPTS

All computers work in more or less the same way, by fetching instructions from their memory one at a time, decoding them, and then executing them. Each instruction is located at a specific address in the computer memory so the first stage in the execution of the program is to fetch the first instruction. This is illustrated in *Figure 2.1*.

During the first phase of an instruction fetch the microprocessor sends the required memory address over the address bus. The address which is selected also receives a simultaneous signal from the control bus to read its contents and place them on the data bus. This is not an instantaneous operation. The memory takes about one tenth of a microsecond to respond. After this time,

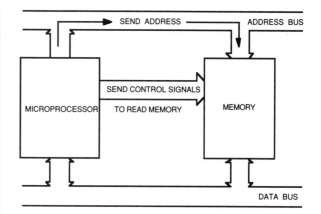

Figure 2.1 Fetching an instruction – step 1

known as the **access time**, the memory places the data from the address on to the data bus. This data is a binary number which is interpreted by the microprocessor as an instruction. This is shown in *Figure 2.2*.

The binary number received by the microprocessor is held in the instruction register. Here the microprocessor begins the work of decoding the instruction. For a typical microprocessor, with an 8-bit data bus, the instruction code could be one of 256 possible codes, so some time is required to undertake the necessary decoding. (There are ways of increasing the number of instructions beyond 256 but they will not be considered here.) The outcome of the decoding phase is that the microprocessor then decides precisely how to execute the instruction.

There are basically two types of instruction:

21

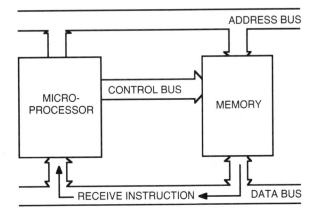

Figure 2.2   Fetching an instruction – step 2

either data, the address of an input/output port, or the address of a memory location in two parts.

A Z80 microprocessor contains certain instructions that require four bytes of memory. These have two parts to the **op-code**. *Figure 2.3* shows that the **op-codes**, the addresses and data are all composed of binary numbers with eight bits. Therefore the microprocessor can mix up the information which is intended as an **op-code** with that intended to be an address. This happens only if the processor is in some way confused because either an instruction is incorrectly formed in the program or it is asked to jump to an address in the middle of an instruction. During normal operation the **op-code** contains information regarding the

(a) Those that require only an internal microprocessor operation.
(b) Those that require extra information from the system.

For instructions that require only an internal microprocessor operation the execution takes place immediately by either an internal data transfer or an arithmetic operation. However, many instructions require additional data or an address to be fetched from the memory.

When an instruction requires an external operation to be carried out, the execute part of the instruction causes another cycle to take place. This may take place more than once if extra information is required. The next address is placed on the address bus and an additional memory read instruction is sent via the control bus. This causes the memory to respond with data that can then be used to complete the instruction which was requested by the microprocessor.

Instruction Format

Any microprocessor has a range of instructions and these may occupy one, two, three or occasionally four bytes of memory. In every case the first byte of memory always contains a code known as the **operation code** or OP-CODE for short. This tells the microprocessor what to do. Typical one, two, three and four byte instructions are illustrated in *Figure 2.3*. The **op-code** is always the first byte but it may also be followed by

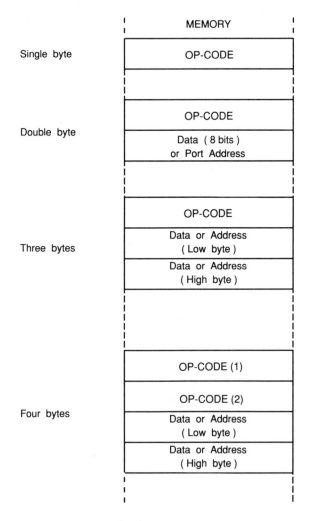

Figure 2.3   Instruction formats

number of bytes required to complete the instruction and in this way the processor is always prepared to receive the next instruction **op-code** when the previous instruction has been completed.

When a computer is first switched on the random access memory contains random numbers in binary patterns, which is generally referred to as 'garbage'. However, since these patterns are in binary, the microprocessor may treat them as though they are instructions. This tends to happen when a program is entered into memory but is not terminated correctly, so that after completing the required program the processor then continues to read the garbage in the memory and tries to treat it as an extension to the program. Clearly this should not be allowed to take place, and therefore it is important that all programs entered into memory are terminated correctly. This may mean that an instruction to halt the processor is used at the end of a program or more likely, an instruction to begin again at the start of a program is introduced.

2.2 COMPUTER TIMING

All microprocessor actions are synchronised by a continuous train of pulses known as a **clock**. These **clock** pulses are generally derived from an oscillator circuit which has a very stable frequency, and is controlled by a quartz crystal. Typically there are about two million clock pulses

per second, i.e. the clock rate is 2 MHz. In modern microprocessors the clock rate is generally higher, so that with some the clock rate will now be as high as 25 MHz or 30 MHz. However, the same basic principles apply and all timing is still derived from the basic cycle of the clock.

In many systems the crystal frequency is actually twice the clock rate, but this is divided by two before it is used as the clock signal.

One cycle of the CPU clock is known as a **T-state** (*Figure 2.4*). If the clock frequency is 2.0 MHz, then each **T-state** lasts only 500 nanoseconds. In general:

$$\text{Time per T-state} = \frac{1}{\text{Clock frequency}}$$

The clock signal is fed into the microprocessor control unit, where it is used to synchronise its operation. Each clock edge is used to trigger some action in either external or internal circuits. Because the control unit is such a complex device, it is controlled by an internal microprogram which runs continuously. One cycle of this microprogram is known as a **machine cycle**. The number of **T-states** in a machine cycle is not fixed, but depends upon the instruction which the microprocessor is executing. The shortest machine cycle for a Z80 has 3 **T-states** while the longest has 6.

The number of machine cycles required to fetch, decode and execute an instruction also depends upon the instruction itself.

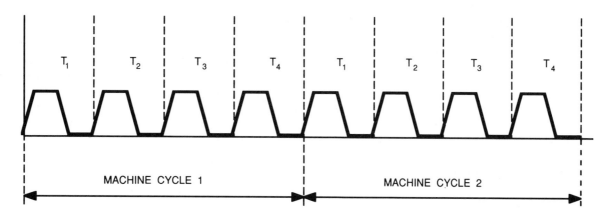

Figure 2.4 Clock cycles

Some instructions require only one machine cycle, while others may require up to six. As soon as the op-code (the first byte of any instruction) is fetched from the memory the information encoded into it will indicate to the control unit how many machine cycles are required and how many **T-states** there will be per machine cycle.

In the Z80 microprocessor, the shortest instructions have one machine cycle of four **T-states**. If the clock rate is 2 MHz they would take 2 microseconds to complete. The longest instructions take six machine cycles and have 23 **T-states**. They would therefore take 11.5 microseconds to complete. It is possible to work out exactly how long a program will take to execute by counting the total number of **T-states** required and multiplying by the time per clock pulse. This is the method used whenever an exact time delay has to be generated by the microprocessor.

The first thing a microprocessor must do with every new instruction is to fetch the op-code from memory. This is a set sequence of events and is known as an **op-code fetch cycle**. During the op-code fetch cycle, the microprocessor goes through the same sequence of events:

- *T-state 1*   The memory address is placed on the address bus and the control signals are issued to read data from memory.

- *T-state 2*   There is a time delay while the memory responds with the requested data, during which time the microprocessor automatically adds 1 to the previously used memory address.

- *T-state 3*   The data from the memory is now transferred back to the CPU over the data bus and the CPU reads it into the **instruction register**.

- *T-state 4*   The CPU decodes the instruction and sets up the necessary control signals to execute it immediately or else in the subsequent machine cycles.

For two byte or three byte instructions the next thing the CPU must do is to read extra data from memory. This operation is very similar to the op-code fetch cycle, except that the data is not placed into the instruction register by the processor. For a three byte instruction this memory read operation must take place twice.

During the execution of the instruction other operations may take place each of which has its own type of cycle in the processor. These include a **memory write**, an **input read**, or an **output write** cycle. In addition there are one or two special cycles that can take place which need not be discussed here.

In summary, the execution of most instructions in a microprocessor is carried out by using a combination of the following microprocessor cycles:

- **memory read.**
- **memory write.**
- **input port read.**
- **output port write.**

Each of these microprocessor operations takes place in a way that is precisely defined by the device manufacturer. Therefore it is possible to predict the signals that will appear at any moment on any of the microprocessor pins and this can be documented in the form of timing diagrams which will be investigated later.

2.3 DETAILED INTERNAL DIAGRAM OF THE Z80

Figure 1.11 (page 11) showed the main parts of which the programmer has to be aware. However, there are a number of additional elements which must be taken into consideration before the execution of an instruction can be completely explained. These are shown in *Figure 2.5*. There are four registers shown in *Figure 2.5* which are 'invisible' to the user.

In the register block the additional registers W and Z are used by the control unit whenever an address has to be stored during the execution of an instruction. They act as normal 8-bit registers which are combined to form one 16-bit register.

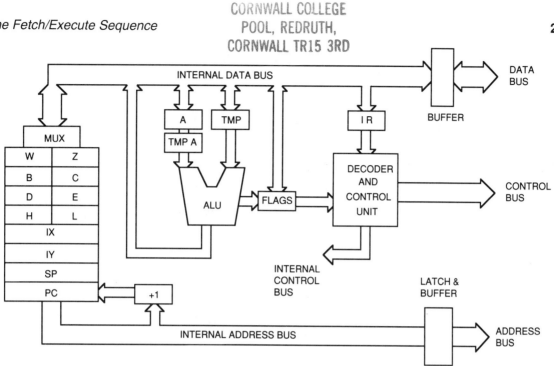

Figure 2.5   Z80 detailed diagram

As with all the other registers, each one is loaded via the internal data bus but as a pair they can be placed upon the internal address bus ready for use by the system.

Register TMP is an 8-bit temporary register used for general data storage. It also holds one of the data bytes prior to its processing in the arithmetic logic unit (ALU). Register TMP A is a temporary accumulator which holds data to be processed in the arithmetic logic unit. Since the result of an ALU calculation is returned to the accumulator A, a temporary accumulator must be included so that the result produced in any calculation cannot immediately feed back into the ALU input and cause an error.

There is also a buffer shown on the data bus. This allows the external and internal section of the data bus to be isolated when required. Often when operations are taking place on the external data bus different operations may be taking place internally and this buffer isolates the two. The address bus is shown with a latch and buffer so that the external and internal address buses can hold different addresses. The latch is a device which is used to hold the address steady on the external address bus even though the internal address bus

may change. In fact this operation takes place during every op-code fetch cycle in **T-state 2**.

2.4 SIMPLE INSTRUCTION EXECUTION

Using the internal diagram of the Z80 microprocessor it is easy to visualise how certain instructions take place. Although the full range of instructions have yet to be introduced, those chosen for the following explanations are relatively simple and their function should be clear.

EXAMPLE 1
Load the accumulator with the number:

n (LD A,n)

This is a two-byte instruction and takes two machine cycles to complete. The complete instruction would occupy two adjacent memory locations. For example, LD A,08 would be 3E 08 in hex code. *Figure 2.6* (pages 26–7) shows the fetch and execute parts of the instruction. Each diagram shows what is happening during a single **T-state**

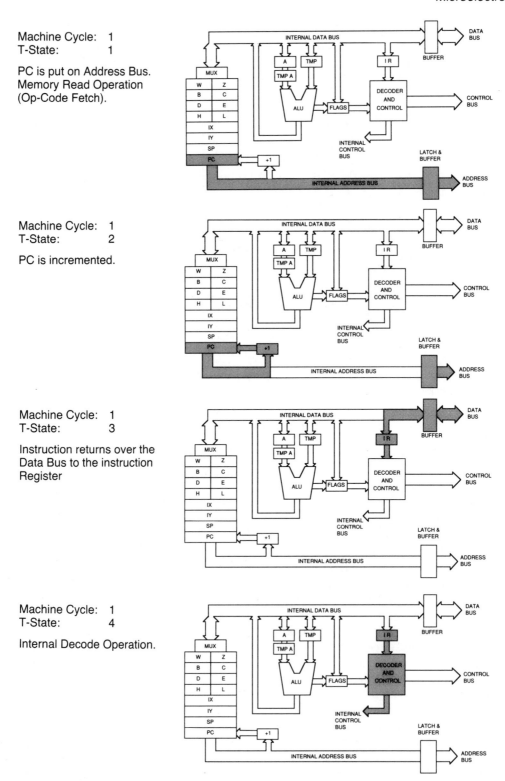

Machine Cycle: 1
T-State: 1

PC is put on Address Bus.
Memory Read Operation
(Op-Code Fetch).

Machine Cycle: 1
T-State: 2

PC is incremented.

Machine Cycle: 1
T-State: 3

Instruction returns over the
Data Bus to the instruction
Register

Machine Cycle: 1
T-State: 4

Internal Decode Operation.

Figure 2.6   LD A,n instruction execution

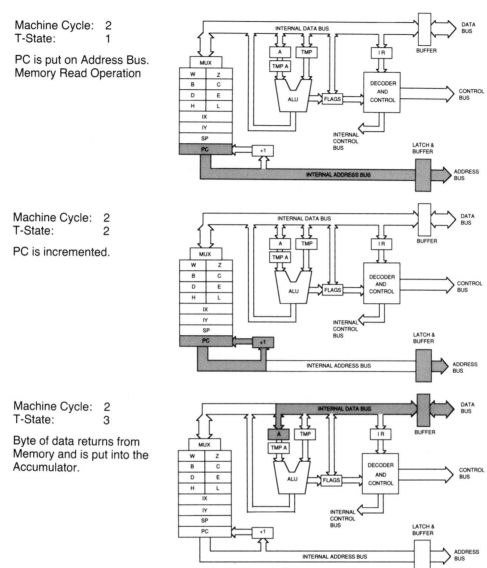

Machine Cycle: 2
T-State: 1

PC is put on Address Bus.
Memory Read Operation

Machine Cycle: 2
T-State: 2

PC is incremented.

Machine Cycle: 2
T-State: 3

Byte of data returns from
Memory and is put into the
Accumulator.

Figure 2.6 *continued*

and the complete instruction takes two machine cycles or seven **T-states** to complete.

EXAMPLE 2

Add data into the accumulator from the memory address whose value is held in the HL register pair (ADD A,(HL)) (*Figure 2.7*, page 28).

This is a single-byte instruction but it takes two machine cycles and part of a third to complete

since its execution requires the processor to read data from memory and add it to the accumulator. Normally, before this instruction is used in a program, the memory address information has to be loaded into the H and L registers.

The first four T-states of this instruction are the same as those given in Example 1 for the LD A,n instruction. They fetch the op-code from the address held in the program counter and return the op-code to the instruction register where it is held.

Machine Cycle: 2
T-State: 1

Contents of the HL pair are
put on Address Bus.
Memory Read Operation

Machine Cycle: 2
T-State: 2

Wait for data from Memory.

Machine Cycle: 2
T-State: 3

Data from Memory is put
into Register TMP. TMP A
is loaded from A.

Machine Cycle: 1
T-State: 1
(next instruction)

PC is put on Address Bus.
Memory Read Operation.
Also, TMP A is added to
TMP and the result put into
A. Flags are affected.

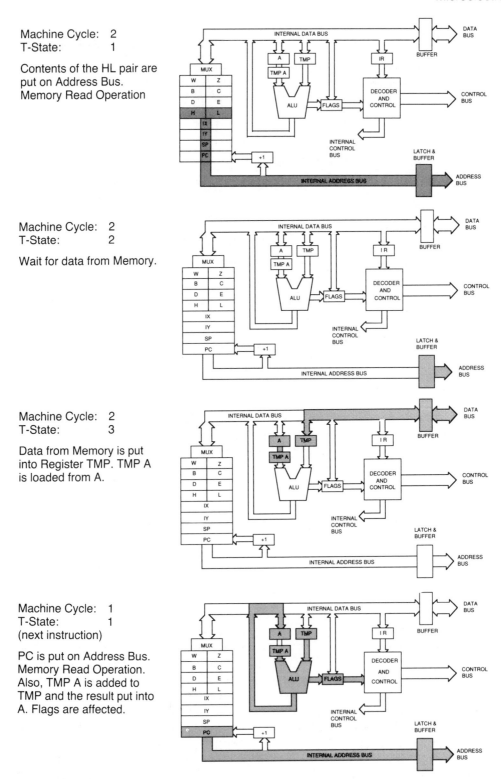

Figure 2.7   ADD A,(HL) instruction execution

EXAMPLE 3

Jump to address nn (JP nn) (*Figure 2.8*).

This is a three byte instruction in which the op-code is followed by the new address, low byte first, high byte last. For example, Jump to address 1234 hex would be C3 34 12 in hex code.

The first four T-states of this instruction are the same as those given in Example 1 for the LD A,n instruction. They fetch the op-code from the address held in the program counter and return the op-code to the instruction register where it is decoded.

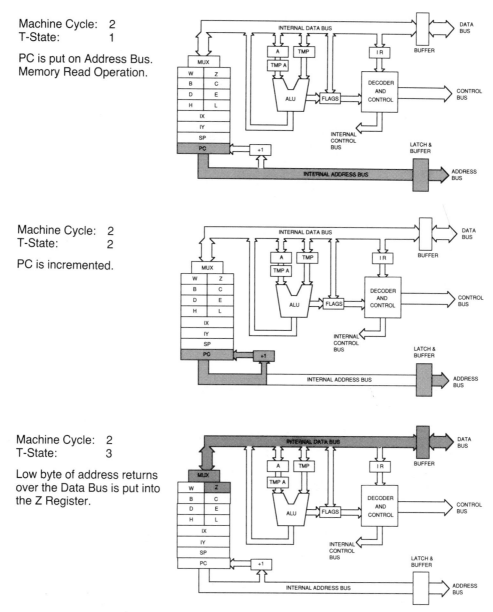

Machine Cycle: 2
T-State: 1

PC is put on Address Bus.
Memory Read Operation.

Machine Cycle: 2
T-State: 2

PC is incremented.

Machine Cycle: 2
T-State: 3

Low byte of address returns over the Data Bus is put into the Z Register.

Figure 2.8   JP nn-instruction execution

Machine Cycle: 3
T-State: 1

PC is put on Address Bus.
Memory Read Operation.

Machine Cycle: 3
T-State: 2

PC is incremented.

Machine Cycle: 3
T-State: 3

High byte of Address re-
turns over the Data Bus is
put into the W Register.

Machine Cycle: 1
T-State: 1

W and Z Register contents
replace the Program
Counter on the Address
Bus. Note: PC is updated
in the next T-State.

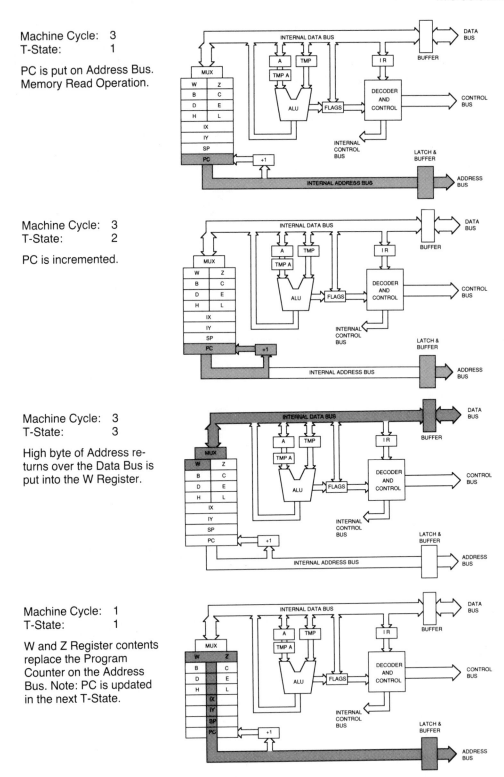

Figure 2.8   *continued*

EXAMPLE 4

Output to port n (OUT (n),A) (*Figure 2.9*).

This is a two-byte instruction which outputs data from the accumulator to port number n. In machine code to output data to say Port 81 hex, the code would be D3 81. However, this instruction requires three machine cycles for its execution, the first two machine cycles read the complete instruction from memory and then the third is required to execute it.

The first four T-states of this instruction are the same as those given in Example 1 for the LD A,n

instruction. They fetch the op-code from the address held in the program counter and return the op-code to the instruction register where it is decoded.

Trying to show all the things that are happening when a microprocessor executes an instruction takes considerable space and it is therefore fairly difficult to describe. However, by producing a timing diagram for each machine cycle it is relatively easy to construct an exact set of timing waveforms which corresponds to the signals which would be found on the individual microprocessor pins.

Machine Cycle: 2
T-State: 1

PC is put on Address Bus.
Memory Read Operation

Machine Cycle: 2
T-State: 2

PC is incremented.

Machine Cycle: 2
T-State: 3

Port Address is put into register Z.

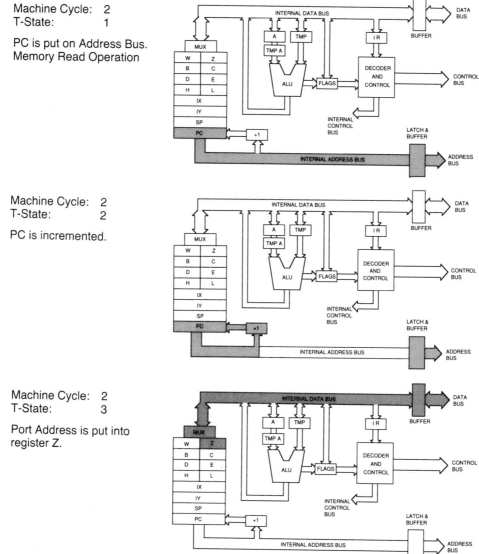

Figure 2.9 OUT (n),A instruction execution

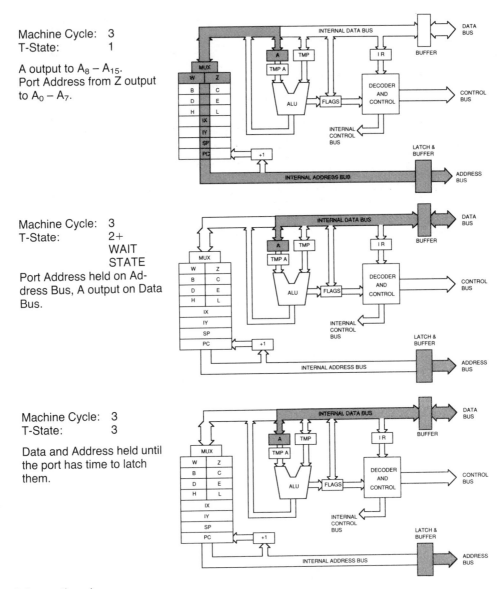

Machine Cycle: 3
T-State: 1

A output to $A_8 - A_{15}$.
Port Address from Z output to $A_0 - A_7$.

Machine Cycle: 3
T-State: 2+
                WAIT
                STATE
Port Address held on Address Bus, A output on Data Bus.

Machine Cycle: 3
T-State: 3

Data and Address held until the port has time to latch them.

Figure 2.9 *continued*

2.5 TIMING WAVEFORMS

The five basic timing waveforms are set out in *Figures 2.10–.14*. When considering how an instruction is executed the waveforms given below are simply put together end to end until the complete instruction is defined.

Instruction Fetch Cycle

A number of things are worth noting in the instruction fetch timing diagram (*Figure 2.10*). The first is that during **T-state** 1 and 2 the contents of the program counter are placed on the address bus so that data can be read from the required memory address. However, during the end of T_3 and T_4 the Z80 places a special address on the address bus which is used to refresh dynamic memory devices if they are used in the system. This is a special feature of the Z80 not found in other microprocessors, and is transparent to the programmer. Notice that two negative going pulses appear on the memory request line, each

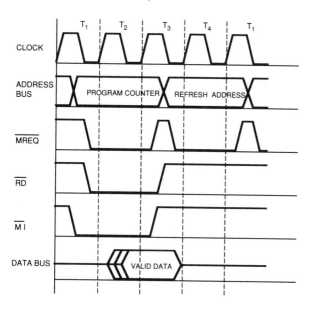

Figure 2.10 Instruction fetch timing waveforms

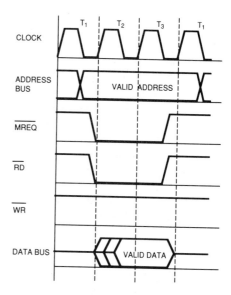

Figure 2.11 Memory read timing waveforms

one corresponding with a valid address on the address bus. Also, the instruction fetch cycle shows the $\overline{MREQ}$, $\overline{RD}$ and $\overline{M1}$ pins all going low simultaneously during T_2. Data from the memory address containing the requested instruction appears on the data bus at a time that depends upon the access time or speed of response of the memory. Therefore this may appear at any time after the beginning of T_2, but it must be present by the time the rising edge of the $\overline{RD}$ pulse occurs, otherwise a read error will occur.

The instruction fetch cycle is the first machine cycle of every instruction.

If an instruction requires extra information from memory then a memory read cycle takes place. This is illustrated in *Figure 2.11* and is very similar to the instruction fetch cycle apart from the fact that the $\overline{M1}$ pulse does not go low during a read cycle.

Note that during the memory read cycle only $\overline{MREQ}$ and $\overline{RD}$ go low. The write signal, $\overline{WR}$ stays at a logical 1. Data returns from the memory after the access time has elapsed, which depends upon the memory chip as before. However, the data must be valid before the rising edge of the $\overline{RD}$ signal. The complete machine cycle is only three **T-states** long.

During the memory write machine cycle the

$\overline{MREQ}$ signal and the $\overline{WR}$ signal are both active, although the $\overline{WR}$ signal is active for a much shorter time. Data is issued by the CPU during **T-state** 1 and this remains valid on the data bus until after the rising edge of the $\overline{WR}$ signal which writes the data into the memory address specified on the address bus (*Figure 2.12*).

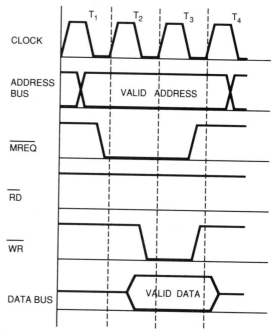

Figure 2.12 Memory write timing waveforms

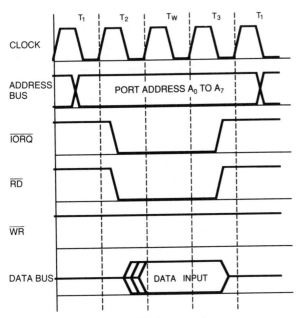

Figure 2.13   Input read timing waveforms

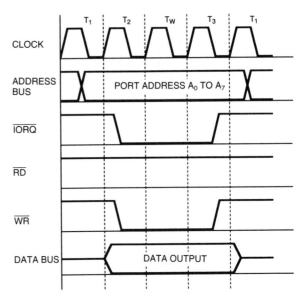

Figure 2.14   Output write timing waveforms

Input read operations (*Figure 2.13*) are very similar to memory read operations apart from one or two slight differences. First, an extra **T-state** is added between T_2 and T_3 for an input/output operation which is known as a **wait state**, T_w. This allows an extra clock cycle to the input operation which gives time for any slow port to respond with its data. The $\overline{IORQ}$ signal is active at logic 0 instead of the $\overline{MREQ}$ signal and also the $\overline{RD}$ signal is low simultaneously for an input operation. The input data must remain valid on the data bus until after the rising edge of the $\overline{RD}$ signal.

The output write cycle (*Figure 2.14*) is almost identical with the input read cycle in that it has an extra wait state T_w added to allow for slow ports to operate. During this cycle the IORQ and WR signals are both active low.

Figures 2.10–.14 show the five basic machine cycles that make up the vast majority of all the microprocessor operations. Instructions use some or all of these machine cycles as necessary so that when two or more machine cycles are required, the T_1 state shown at the end of each diagram simply becomes the first **T-state** of the next machine cycle. The diagrams are effectively added end to end as required. The activity on the various control lines can be summarised for each of the cycles as in *Figure 2.15*.

Machine Cycle	$\overline{MI}$	$\overline{WR}$	$\overline{RD}$	$\overline{IORQ}$	$\overline{MREQ}$
Instruction Fetch	●		●		●
Memory Read			●		●
Memory Write		●			●
Input Read			●	●	
Output Write		●		●	

Figure 2.15   Machine cycle control signal summary

2.6 INTERPRETATION OF TIMING DIAGRAMS

Case Study – The Memory Read Cycle

The previous timing diagrams have shown the type of waveforms that would be expected to be seen on the system buses during the five main cycles which the processor executes. These cycles are:

- Instruction fetch.
- Memory read.
- Memory write.
- Input read.
- Output write.

It has already been suggested that everything within the microprocessor system is very accurately timed and synchronised by the crystal oscillator known as the **clock**. Precisely how this is done can be seen better by examining one of the machine cycles in greater detail. The machine cycle chosen for this exercise is the **memory read cycle** since it is relatively simple and also is used the most frequently after the instruction fetch cycle. Other machine cycles would have very similar timing constraints, and the principles described here are applicable to all the other machine cycles.

Figure 2.16 shows the complete timing diagram similar to that which would be found in the manufacturer's manual for the microprocessor chip.

The **clock** waveform for the memory read cycle normally has three **T-states**, but in any cycle the number of **T-states** may be extended indefinitely by adding extra **wait** states labelled T_w. Here, only one **T-state** has been added to show the effect on the waveforms when this occurs. At either side of

the **wait** state broken lines indicate that these may go on indefinitely.

All the timing is synchronised from the **clock**, and it appears that the sides of the clock pulse and all the other pulses are slightly sloped. This is to indicate that although the **clock** appears to be a squarewave, the rise and fall times are significant at the speed at which the **clock** is operating, typically between 1 and 6 MHz. Events are triggered when the **clock** either reaches the logic 1 level or logic 0 level, and these points are marked by small dashes on the **clock** waveform.

All the timings below refer to a **clock** speed of 4 MHz for the Z80.

Time (1) – CLOCK rising edge to a valid address

After the **clock** pulse has reached its logic 1 state a time no more than 110 ns must elapse before a valid memory address is placed on the address bus. During this cycle the address is generally the contents of the program counter or that of a register pair. Since the address appears

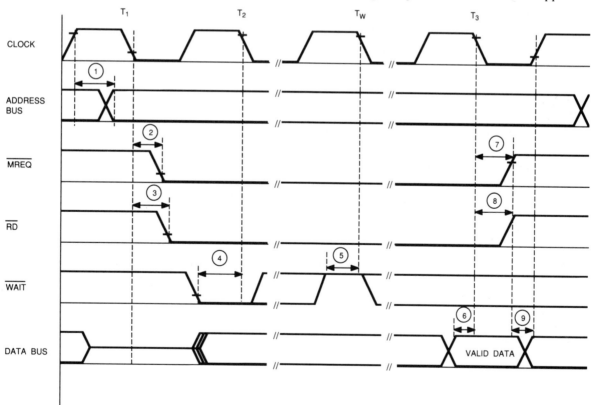

Figure 2.16 Detailed timing diagram – memory read

relatively early in this cycle on the address bus this gives memory devices plenty of time to decode the address ready to provide the data from it.

Time (2) – CLOCK falling edge to $\overline{MREQ}$ delay As soon as the **clock** falling edge has occurred there is a maximum delay of 70 ns, before the memory request signal goes low. This signal is often used to activate logic devices that help to decode the address on the address bus.

Time (3) – Almost at the same time as the $\overline{MREQ}$ signal is going low, the $\overline{RD}$ signal also goes low The maximum time allowed here is 80 ns. As soon as both $\overline{MREQ}$ and $\overline{RD}$ are both low, together with a valid address on the address bus, memory devices have all the signals they need to be able to respond with the data from the address selected. The time taken from the falling edge of the $\overline{RD}$ signal until valid data appears during **T-state** 3 is known as the **access time** of the memory. Since this time is not a direct function of the CPU timing it is not shown on the diagram explicitly although it can be seen indirectly.

Time (4) – WAIT state generation The **wait** line is an input to the CPU which can force it to add extra **T-states** to each machine cycle. It is ignored during **T-state** 1 but if the **wait** line is low at a time when the trailing edge of the **clock** in **T-state** 2 occurs then the CPU adds an extra **wait** state to the machine cycle. The **wait** line must go low at a predetermined time before the negative going edge of the **clock** in T_2. Typically this is a minimum of 60 ns.

Time (5) – WAIT state set-up time During each **wait** state the logic level of the **wait** line is sampled. If this has been changed to a logic 1 at least 60 ns before the negative going edge of the **clock** then the **wait** state terminates at the end of that **clock** cycle and the system resumes with the final **T-state**, T_3 of the machine cycle. This is the normal course of events and unless very slow memory devices are being used or there is some other reason for adding **wait** states, generally most machine cycles omit them.

Time (6) – Data set-up time The memory must have responded with valid data by the time T_3 arrives. It must be valid on the data bus at least 40 ns before the **clock** in T_3 reaches its logic 0 level. This gives the system time to respond by reading the data a few nanoseconds later. During the previous **T-states** the data on the data bus could either be 1s or 0s and is ignored by the CPU.

Time (7) – CLOCK trailing edge to $\overline{MREQ}$ rising edge About 70 ns after the negative going edge of the CLOCK in T_3 the $\overline{MREQ}$ signal goes back to its high state. This signifies the end of this **memory read** cycle.

Time (8) – CLOCK negative edge to RD rising edge No more than 70 ns after the falling **clock** edge in T_3, the $\overline{RD}$ pulse also goes back to a logic 1. It is this transition that actually **reads** the **data** from the data bus into the appropriate **register** in the CPU and therefore the **data** on the bus must be valid at this time.

Time (9) – Valid data delay **Data** usually stays valid on the data bus for a short time after the RD signal goes high because of the time taken for memory devices to close down their output buffers. However, there is no requirement for any delay and therefore Time (9) can be zero.

The preceding explanation of the timing of the CPU cycle may seem particularly complex. However, notice that everything happens entirely automatically and it is only when systems are being designed that the precise timing of cycles needs to be considered. In any system that has been working for some time the only reason for examining the waveforms may be to discover a fault within the system and it is therefore useful to know which events should happen and in which order on the buses.

If an oscilloscope is used to examine waveforms in the system, it is particularly difficult to examine more than one or two signals at any moment simply because of the nature of the instrument. Some modern oscilloscopes provide multitrace facilities whereby up to 8, or even 16, traces can be viewed simultaneously. However, modern logic devices known as **logic analysers** provide the best means of examining the data, address and control bus contents since they allow up to 32 simultaneous signals to be examined and the contents to be displayed on the oscilloscope or VDU later. This provides the simplest way of examining signals in a microprocessor system.

Summary

This chapter has described the basic fetch execute cycle of the microprocessor. The most important points are:

- The first byte of each instruction contains the op-code.
- The op-code is fetched during the first machine cycle of each instruction.
- When the op-code is decoded, the microprocessor then executes the instruction by implementing the required machine cycles.

- The main microprocessor machine cycles are:
  (a) Instruction fetch.
  (b) Memory read.
  (c) Memory write.
  (d) Input read.
  (e) Output write.
- Microprocessor timing is controlled by a **clock** waveform.
- Each clock cycle is known as a 'T-state'.
- A machine cycle is made up of 3–6 T-states.
- An instruction may require up to 6 machine cycles to complete its execution.
- Each machine cycle has a predefined sequence of signals associated with it which can be identified on the system buses.
- The data present in a manufacturer's timing diagram allows the bus activity to be accurately predicted for any instruction.

Questions

2.1 Briefly describe the part played by the control bus in the fetch/execute cycle of a microprocessor.

2.2 What is the function of an op-code?

2.3 What is the function of a clock in a microprocessor system?

2.4 Define the terms:

(a) T-state.

(b) Machine cycle.

2.5 What determines the number of T-states and machine cycles an instruction will require for completion?

2.6 Briefly describe how a microprocessor keeps track of how many bytes constitute each instruction.

2.7 What machine cycles would be required when the instruction LD (1950H),A (Load address 1950H from the accumulator) was executed?

2.8 Which signals are active during the output write cycle in a Z80 microprocessor?

2.9 Why is there a time delay between the address and control signals being applied to a memory device, and the data becoming available?

The microprocessor instruction set

When you have completed this chapter, you should be able to:

1. *Understand that the instructions recognised by microprocessors are known as the **instruction set**.*

2. *Group the main microprocessor instructions into the following sets:*
   (a)   Data transfer.
   (b)   Data manipulation.
   (c)   Flow of control.

3. *Appreciate the differences between assembly language and machine code instruction formats.*

4. *Appreciate the steps required in the production of a simple machine code program.*

5. *Understand the need for different addressing modes.*

6. *Understand the concepts of binary number representation in a computer.*

7. *Perform basic binary mathematical operations.*

8. *Understand the function of the **flag** register.*

9. *Understand the use of arithmetic and logic instructions in simple programs.*

10. *Appreciate the uses of BCD number representation and ASCII code.*

11. *Appreciate that instructions can be used to change the flow of a computer program such as conditional and unconditional jumps.*

3.1 THE INSTRUCTION SET AND PROGRAMMING

When a new microprocessor is designed the manufacturer must make a number of significant decisions. These include for example, the width of the data bus, the size of the address bus, the number of registers, and the range of instructions to which the microprocessor will respond. The range of instructions decided upon by the manufacturer is known as the **instruction set** of the microprocessor. Different manufacturers go about the design in different ways and therefore the range of instructions which can be understood by any processor may be considerably different to those understood by another processor. Also, some processors can respond to far more instructions than others.

For example, compare the Z80 and the 6502 microprocessors, both of which were designed about the same time. The Z80 microprocessor has 696 separate instructions, whereas the 6502 has only 151. This means that a lot of functions which can be performed using a single Z80 instruction require several 6502 instructions. Also, with an 8-bit data bus there are only 256 different combinations which can be used for an op-code. Therefore some of the Z80 instructions require two bytes for the op-code which means that they take longer to execute than the single byte instructions of the 6502.

In the early days of microprocessor design,

there was a tendency towards designing more and more complex microprocessors with each succeeding generation so the 16- and 32-bit processors which were designed were capable of executing a far wider range of instructions than the earlier processors. This also led to a large number of instructions requiring a very long time to execute them. Some of the latest thinking in microprocessor design is to reduce the number of instructions which the microprocessor can execute. This creates a processor known as a **reduced instruction set computer** (RISC) which can actually execute instructions faster than those processors that have been designed to have large and complicated instruction sets. In a RISC machine, whenever complicated instructions are required, such as a binary multiplication, then a short program made up of faster instructions has to be executed in order to carry out this function. The Z80 is a product of an earlier line of thinking, and its instruction set therefore contains a range of both simple and more complicated instructions.

Throughout this book, the instruction set, the programs and the examples are all based on the Z80 microprocessor. However, since this processor was developed from the Intel 8080, the majority of the instructions described are machine-code compatible with this and the 8085 processor. Unfortunately the codes used in assembly language are not compatible, so that the assembly language written for the 8080 could not be used with a Z80 assembler even though the resulting machine code may run on a Z80 microprocessor.

Fortunately, all microprocessor instructions can generally be classified into one of three main groups which are common to all types of processor. These are:

(a) Data transfer group.
(b) Data manipulation group.
(c) Flow of control group.

These groups of instructions make up the complete instruction set, and individual instructions fall within one of these groups. Whatever the microprocessor, the job of the programmer is to select from the range of instructions available, the instructions best suited to carry out the task that

has to be performed. To do this, the programmer must be familiar with the whole range of instructions that the microprocessor can use and carefully select those that are required for each function. In many circumstances, there are two or three different ways to perform this same task and therefore the skill and experience of the programmer plays a vital role in the efficiency of the programs produced.

3.2 ASSEMBLY LANGUAGE AND MACHINE CODE

The only language that the computer understands is **binary code**. Whatever language the programmer chooses to write programs in, there must always be some means of translating that program into the binary code which the computer requires, otherwise it will never be able to run. However, as far as the programmer is concerned binary code is a particularly difficult method of programming a computer, although it was the only method used in the very early days of computing. Nowadays, computer programs are written in one of two ways, using either:

(a) **A high level language**, or
(b) **A low level language**.

High level languages – such as BASIC, COBOL, PASCAL etc. – are generally used whenever large programs are required and there are no restrictions in terms of computer memory or program execution time. Programming in a high level language is a lot easier than programming in a low level language, but it does require a system with more memory and the completed programs tends to operate more slowly than their equivalent which have been programmed using a low level language. Generally, high level languages are inappropriate for small microprocessor based systems or those that have to operate at the highest possible speed.

Low level languages are noted for their low memory requirements in a system, and their high operating speed. The problem with this approach, however, is that programs are more difficult to write in low level languages and also they are not transportable between different microprocessors.

If this was a requirement of a system a completely new program would have to be written in most cases. The main advantage of using a low level language is that it allows the programmer direct interaction with the microprocessor chip at its register level.

Microprocessor instructions could be specified in one of four ways:

(a) *English language representation* – such as, load the accumulator with data from input port 80H.
(b) *Assembly language* – such as, IN A,(80H)
(c) *Hexadecimal machine code* – such as, DB 80.
(d) Binary machine code – such as, 1 1 0 1 1 0 1 1 1 0 0 0 0 0 0 0.

Clearly it would be impracticable to express computer programs in English form, simply because they would take far too long to write out and there could be misinterpretation of the program due to the type of language used. Different programmers may express the same concept in different ways which would be very confusing. Some simpler form of computer program representation must be used.

The next best method of writing programs is to use the assembly language coded form of instructions, known as **mnemonics**. A mnemonic is simply a coded way of writing down computer instructions which makes it easy to remember, and it is important to understand this form of writing programs. Each instruction can be represented by a simple two or three letter command followed by some variables known as the **operands**. Since the assembly language mnemonic, known as the **op-code**, is related to the English language representation of an instruction it means that they are very easy to remember. However, before the assembly language mnemonics can be used in a computer, each one must be translated into its correct binary code. The process by which this is achieved depends upon the facilities available to the programmer.

Hexadecimal machine code is also frequently used to represent computer instructions. Hex code is simply a convenient way of writing the binary information that the computer will require to operate. By using hexadecimal codes, this makes the task of understanding the binary a lot

easier for the programmer. It is particularly important to understand the relationship between hexadecimal codes and their associated assembly language mnemonics.

Binary machine code is only ever used by the computer, or in circumstances where it is important to understand what is happening to an individual bit, for example at an input or an output port, or during logical operations in the microprocessor. In general, hexadecimal codes are sufficient for the programmer; and binary code, although important in certain circumstances, is generally ignored.

Assembly Language

Assembly language is by far the easiest way to write computer programs, largely because the detail of the addresses used in the program, and the hexadecimal codes of the instructions can be largely ignored. This allows the programmer to concentrate on the logical flow of the program and ensure that the program structure is correct before it is translated into machine code. In addition when writing in assembly language, symbols are used for addresses, and **labels** can be used where points in a program need to be indicated. Comments can be added to various points in the program to remind the programmer of what the program is intended to be doing. *Figure 3.1* shows a typical assembly language program. This illustrates several important features. First, there are no addresses in the assembly language program. The labels **start** and **loop** both refer to addresses but at the time of writing the program, these addresses are not known so symbolic addresses or labels are used. The ORG statement refers to the **origin** address, which will be the first address to be used when the program is translated into machine code. The program consists of the mnemonics for the instructions which are listed one after another in the order in which they will be executed. After certain lines of the program a semi-colon precedes a comment. These comments are used by the programmer to indicate what is happening on each line of the program. Notice that they are not directly related to the instruction but indicate some other feature of the

Labels	Mnemonic	Comment
	ORG 1800H	; ORIGIN ADDRESS
START:	LD A, 3FH	
	OUT (83H),A	; INITIALISE OUTPUT PORT
	LD A, 80H	
	OUT (81H), A	; TURN MOTOR CLOCKWISE
	LD BC, 0000 H	; SET UP DELAY COUNT
LOOP:	DEC BC	
	LD A, B	
	OR C	
	JP NZ, LOOP	
		; LOOP UNTIL BC=0

Figure 3.1 Typical assembly language program

program. Comments are optional so that some lines do not require them.

The use of the assembly language programming technique allows the structure of the program to be clearly seen; but before it can be used in a computer, it somehow has to be translated to machine code.

3.3 PRODUCING A MACHINE CODE PROGRAM

An assembly language program can be translated into machine code in one of two ways, depending upon the sophistication of the computer being used. In more advanced computers, having an **assembler** program, the computer can be made to do the translation. This means that the programmer can enter the assembly language mnemonics directly into the computer, and the computer will work out the correct machine code to use for each instruction.

Less sophisticated computers do not have this **assembler** program facility, and therefore the programmer has to perform the translation from assembly language mnemonics to machine code by hand. Fortunately this is a relatively simple process which can be carried out quickly with the aid of suitable tables.

Using an Assembler Program

Computers designed to be used with direct entry of assembly language mnemonics must have a number of features. These are:

(a) A full alphanumeric keyboard.
(b) A display for alphanumeric information.

(c) An **editor** program.
(d) An **assembler** program.
(e) A **monitor** program.

Computers do not have to be particularly large to incorporate these features, and **assembler** programs can be bought for most computers with full keyboard and VDU displays. However, they may also be built into smaller single board computers. A process of producing a machine code program, starting from the assembly language, is shown in *Figure 3.2* (page 42).

Step 1 is done on paper and it is generally not until the complete program has been written that anything is entered into the computer.

An **editor** program is used in the computer to accept the assembly language mnemonics and store them in the computer memory. The **editor** may also allow corrections to be made to the mnemonics if they have been entered incorrectly or if they have to be changed later. On sophisticated computer systems, a **wordprocessor** program may be used instead of a simple **editor**.

As soon as the program has been entered correctly and checked, the **assembler** program is run in Step 3, and this produces the machine code required. In sophisticated assembly language systems, blocks of machine code that have been written and assembled independently may be linked together to form larger programs. However, this step is not always necessary. Once the program has been assembled correctly, with no errors, a **monitor** is used to execute the program. This consists of a program resident within the computer which allows the user to run programs when they are stored in the system memory.

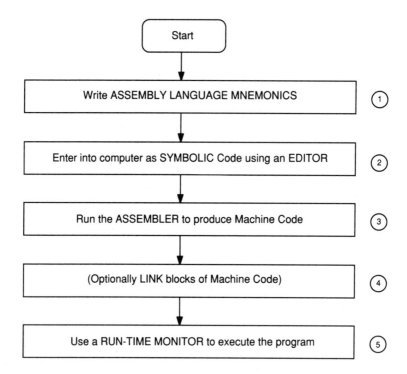

Figure 3.2 Producing a machine code program

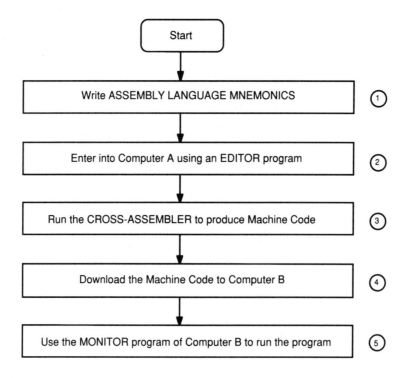

Figure 3.3 Using a cross assembler

Notice that to write a program, assemble it and run it, the computer must contain three other programs:

(a) The **editor**.
(b) The **assembler**.
(c) The **monitor** program.

While it is convenient if all three of these programs are resident within the computer that will run the final program, it is by no means necessary that they must all be present in that machine. For example, it is possible to use the **editor** and **assembler** of one computer and then the **monitor** program of another. In this case the **assembler** is referred to as a **cross-assembler**. *Figure 3.3* shows this method of operation.

The figure shows the **editor** program and **cross-assembler** being used in computer A. This could be any computer, such as an IBM compatible machine, which is capable of running say, a Z80 assembler program. This assembler program is a **cross-assembler**, because the native microprocessor of the machine in which the assembly is taking place is not a Z80. This means that the microprocessor in computer A is incapable of running the machine code produced. Therefore it has to be downloaded into a Z80 based machine before it can be executed, and this is the reason for the download process in Step 4 of the sequence of operations. Once the program has been downloaded then the **monitor** in the target computer can be used to execute the program.

With both of the methods just described, everything works fine as long as there are no mistakes in the program, and it executes perfectly first time. Unfortunately, the chances of this happening, particularly with novice programmers, will be fairly remote. Therefore, there needs to be some method of correcting the programs once mistakes are discovered. In both cases this requires returning to Step 1 and modifying the assembly language mnemonics on paper before any corrections can be made. Once the program has been corrected on paper the whole process has to be repeated with the corrected version of the program until a satisfactory solution is achieved.

Hand Translation

Most people learning microelectronics start with 'hand translation' of assembly language. All of the assembly language must be translated into hexadecimal codes before the programs can be executed. This is not a particularly difficult task and need only take a few moments for each program once use of the code tables has been understood. *Figure 3.4* shows the process diagrammatically.

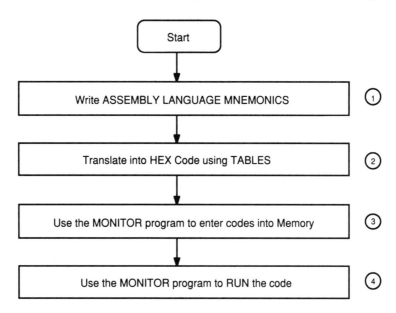

Figure 3.4 The hand assembly translation process

Address	Hex Code	Mnemonic	Comment
			ORG 1800H ; ORIGIN ADDRESS
1800	3E 3F	START:	LD A, 3FH ; INITIALISE OUTPUT PORT
1802	D3 83		OUT (83H), A
1804	3E 80		LD A, 80H ; TURN MOTOR ON
1806	D3 81		OUT (81H), A ; SET UP DELAY COUNT
1808	01 00 00		LD BC, 0000H
180B	0B	LOOP:	DEC BC
180C	78		LD A, B
180D	B1		OR C ; LOOP UNTIL BC=0
180E	C2 0B 18		JP NZ, LOOP
			etc.

Figure 3.5 Typical machine code program

The hand assembly process begins with the same assembly language mnemonics used in each of the other cases. However, the second step in this process is to take the assembly language and add to it both the addresses that will be used and the correct hexadecimal codes for each instruction. *Figure 3.5* shows an example of this, using the same program as in *Figure 3.1*. When the hex codes have been worked out, the monitor program of the computer is used to enter the hex codes directly into memory. Fortunately most monitors will increment memory addresses so that as each pair of digits is entered the monitor program automatically steps on to the next address ready for the next pair of digits. As soon as a program is complete, the monitor program is used to execute it as in the other case. As before if errors are found, then the programmer must go back to the assembly language mnemonics in Step 1 to correct the error before going through the whole process of entering the corrections and trying the program again.

The right-hand half of the machine code program looks exactly like the assembly language program written previously. However, each line of the program has been written in hex code by looking up the correct codes for the mnemonics used in a set of tables. The left-hand column of the machine code program has also been added, and this represents the address of the first byte of the hex code for each instruction. Therefore, the addresses do not follow sequentially but their values reflect the number of bytes used in each instruction.

3.4 INSTRUCTION FORMAT

Most of the instructions within a microprocessor instruction set are designed to operate upon data found somewhere in the system. It may be located in one of the CPU registers, in a memory address, or at an input port. Therefore within each instruction there must be a means of identifying the location of the data to be operated upon. This implies that there must be two parts to each instruction: first, the **operation code**, generally shortened to 'op-code' and second, the address location code, which is known as the **operand**. This is illustrated in *Figure 3.6*.

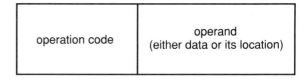

operation code	operand (either data or its location)

Figure 3.6 Typical microprocessor instruction format

The **operation code** can be as few as two bits, or it may be as many as sixteen bits. Similarly, the **operand** may consist of a 3-bit address, such as when registers are used, or it may consist of a full 16-bit address in memory. Therefore the total code for any operation may be as little as one byte (8 bits) or it could be as large as 4 bytes (32 bits). This is illustrated by two examples below.

The first example (*Figure 3.7*) shows a single byte instruction which is used to load the accumulator from the B register. The 8 bits are divided into three parts; the first specifying the operation

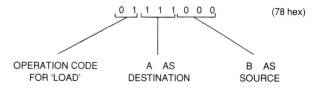

Figure 3.7 Composition of LD A,B instruction

code and the second and third specifying the A and B registers respectively.

The second example shows another **load** instruction, load A from address 1234 hex, but here the complete address is specified since it is a 'Load the accumulator from memory' instruction. The op-code requires 8 bits and is hex code 3A. After that, the complete address has to be specified and this is written using the lower two hexadecimal digits first then the higher two hexadecimal digits. *Figure 3.8* shows the complete breakdown of the instruction.

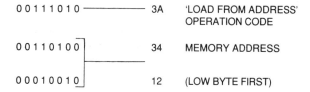

Figure 3.8 Composition of LD A,(1234H) instruction

Notice that although the accumulator is not specified specifically in the hex code format, it is implied by virtue of the op-code used.

Most instructions manipulate data which is found somewhere in the microprocessor system. Generally, this would be in a memory location which has a specific address, or it may be in one of the microprocessor registers, which can also be specified with an address (3 bits). Various methods exist which allow the system to locate the data to be operated upon and these methods are known as the microprocessor '**addressing modes**'.

3.5 ADDRESSING MODES

An addressing mode may be defined as 'the method by which the microprocessor locates the data or address to be used in an instruction'.

All microprocessors have a variety of ways in which they find the data, for example the Z80 has seven addressing modes, whereas the 6502 microprocessor has ten. A large number of addressing modes allows greater flexibility on the part of the programmer in how to use the microprocessor instructions effectively. For example some addressing modes may allow data to be accessed directly, whereas others allow the address of data to be calculated as part of an expression. The main addressing modes used by the Z80 microprocessor as defined below.

Register Addressing

Register addressing means that the data to be operated upon will be found in a CPU register. For example:

LD A,B – Load the data from the B register into the A register (accumulator).

Here, an instruction is required which simply puts the data from Register B into the accumulator. Since the data transfer is between registers within the CPU, it is said to employ 'register addressing'. The addresses of the registers to be used are included in the machine code for the instruction.

Immediate Addressing

Immediate addressing means that the data to be used is found in the memory address immediately after the op-code. For example:

LD A,25H – Load the A register with 25 hex.

This instruction requires two memory addresses, 3E is the op-code in the first address, and 25 is the data in the address immediately afterwards.

This type of instruction, using immediate addressing is always used to load data directly into one of the CPU registers.

Indirect Addressing

Indirect addressing means that the data to be used is found in a memory location whose address is held in a register pair. For example:

LD A,(HL) – Load the accumulator with the data found in the memory location whose value is in the HL register pair.

DESTINATION		IMPLIED		REGISTER							REG INDIRECT			INDEXED		EXT. ADDR.	IMME.
		I	R	A	B	C	D	E	H	L	(HL)	(BC)	(DE)	(IX+d)	(IY+d)	(nn)	n
REGISTER	A	ED 57	ED 5F	7F	78	79	7A	7B	7C	7D	7E	0A	1A	DD 7E d	FD 7E d	3A n n	3E n
	B			47	40	41	42	43	44	45	46			DD 46 d	FD 46 d		06 n
	C			4F	48	49	4A	4B	4C	4D	4E			DD 4E d	FD 4E d		0E n
	D			57	50	51	52	53	54	55	56			DD 56 d	FD 56 d		16 n
	E			5F	58	59	5A	5B	5C	5D	5E			DD 5E d	FD 5E d		1E n
	H			67	60	61	62	63	64	65	66			DD 66 d	FD 66 d		26 n
	L			6F	68	69	6A	6B	6C	6D	6E			DD 6E d	FD 6E d		2E n
REG. INDIRECT	(HL)			77	70	71	72	73	74	75							36 n
	(BC)			02													
	(DE)			12													
INDEXED	(IX+d)			DD 77 d	DD 70 d	DD 71 d	DD 72 d	DD 73 d	DD 74 d	DD 75 d							DD 38 d n
	(IY+d)			FD 77 d	FD 70 d	FD 71 d	FD 72 d	FD 73 d	FD 74 d	FD 75 d							FD 36 d n
EXT. ADDR.	(nn)			32 n n													
IMPLIED	I			ED 47													
	R			ED 4F													

Figure 3.9   Instructions using register addressing

In this case the mnemonic is written with B as the destination and the data 'n' as the source. Hence the mnemonic:

LD B,36H

To work out the correct hexadecimal code, use the final column of the chart, highlighted in *Figure 3.10* to find the row with the destination register B. This gives the code as 06 n, where n is the data to be loaded. Therefore the full instruction is:

LD B,36H – hex code 06 36

Similarly:

Load the Accumulator with AB Hex   LD A,ABH
Hex Code   3E AB

Register indirect instructions   The instructions that use register indirect addresing allow registers to be loaded with the data from memory addresses or vice versa, without having to specify

SOURCE

		IMPLIED		REGISTER							REG INDIRECT			INDEXED		EXT. ADDR.	IMME.
		I	R	A	B	C	D	E	H	L	(HL)	(BC)	(DE)	(IX+d)	(IY+d)	(nn)	n
REGISTER	A	ED 57	ED 5F	7F	78	79	7A	7B	7C	7D	7E	0A	1A	DD 7E d	FD 7E d	3A n n	3E n
	B			47	40	41	42	43	44	45	46			DD 46 d	FD 46 d		06 n
	C			4F	48	49	4A	4B	4C	4D	4E			DD 4E d	FD 4E d		0E n
	D			57	50	51	52	53	54	55	56			DD 56 d	FD 56 d		16 n
	E			5F	58	59	5A	5B	5C	5D	5E			DD 5E d	FD 5E d		1E n
	H			67	60	61	62	63	64	65	66			DD 66 d	FD 66 d		26 n
	L			6F	68	69	6A	6B	6C	6D	6E			DD 6E d	FD 6E d		2E n
REG. INDIRECT	(HL)			77	70	71	72	73	74	75							36 n
	(BC)			02													
	(DE)			12													
INDEXED	(IX+d)			DD 77 d	DD 70 d	DD 71 d	DD 72 d	DD 73 d	DD 74 d	DD 75 d							DD 38 d n
	(IY+d)			FD 77 d	FD 70 d	FD 71 d	FD 72 d	FD 73 d	FD 74 d	FD 75 d							FD 36 d n
EXT. ADDR.	(nn)			32 n n													
IMPLIED	I			ED 47													
	R			ED 4F													

DESTINATION (row label for table)

Figure 3.10 Instructions using immediate addressing

the address as part of the instruction. Instead the memory address must previously have been loaded into a register pair in the CPU. The vast majority of these instructions use the HL register pair which, incidentally, has the letters H and L in its name to represent high and low parts of the memory address. However, in certain circumstances the BC pair and the DE pair can also be used for indirect instructions if the accumulator is also involved in the data transfer.

Remember with all of these instructions, they will not work unless the register pair used to hold the address has previously been loaded with the address to use. One way of loading the memory address is to use the immediate instructions such as LD H,n_1 and LD L,n_2, where n_1 and n_2 represent the parts of the memory address to be used. Notice in *Figure 3.11*, page 50 that the names of the registers used to hold memory addresses are enclosed in brackets. When this is the case, the

DESTINATION			IMPLIED		REGISTER							REG INDIRECT			INDEXED		EXT. ADDR.	IMME.
			I	R	A	B	C	D	E	H	L	(HL)	(BC)	(DE)	(IX+d)	(IY+d)	(nn)	n
REGISTER	A		ED 57	ED 5F	7F	78	79	7A	7B	7C	7D	7E	0A	1A	DD 7E d	FD 7E d	3A n n	3E n
	B				47	40	41	42	43	44	45	46			DD 46 d	FD 46 d		06 n
	C				4F	48	49	4A	4B	4C	4D	4E			DD 4E d	FD 4E d		0E n
	D				57	50	51	52	53	54	55	56			DD 56 d	FD 56 d		16 n
	E				5F	58	59	5A	5B	5C	5D	5E			DD 5E d	FD 5E d		1E n
	H				67	60	61	62	63	64	65	66			DD 66 d	FD 66 d		26 n
	L				6F	68	69	6A	6B	6C	6D	6E			DD 6E d	FD 6E d		2E n
REG. INDIRECT	(HL)				77	70	71	72	73	74	75							36 n
	(BC)				02													
	(DE)				12													
INDEXED	(IX+d)				DD 77 d	DD 70 d	DD 71 d	DD 72 d	DD 73 d	DD 74 d	DD 75 d							DD 38 d n
	(IY+d)				FD 77 d	FD 70 d	FD 71 d	FD 72 d	FD 73 d	FD 74 d	FD 75 d							FD 36 d n
EXT. ADDR.	(nn)				32 n n													
IMPLIED	I				ED 47													
	R				ED 4F													

Figure 3.11   Instructions using the register indirect addressing

value in the register pair, is said to **point to** the memory address. For example:

   LD A,(HL) can be read as   Load the accumulator with data from the memory address pointed to by the contents of the HL register pair.

So, to load the data from address 1234H into the accumulator, the following instructions could be used.

LD H,12H
LD L,34H
LD A,(HL)

This may seem a rather long way of getting a value from a memory address, but the H and L registers may be loaded in one instruction, which will simplify the operation and also once the address is loaded it can be used repeatedly or modified for later instructions.

One more example:

Load the E Register contents into Memory Address 1820H.
LD H,18H
LD L,20H
LD (HL),E

Extended addressing   Occasionally there is a need to load the accumulator directly from a memory address, without wanting to load the HL register pair first. Generally this is where a single byte of data is required from a specific memory address. Similarly, there is sometimes a need to load data from the accumulator directly into a memory address. In these cases the extended addressing instruction (*Figure 3.12*) is used which occupies 3 bytes, the last two of which contain the memory address to be used. Note that whenever

			IMPLIED		REGISTER							REG INDIRECT			INDEXED		EXT. ADDR.	IMME.
			I	R	A	B	C	D	E	H	L	(HL)	(BC)	(DE)	(IX + d)	(IY + d)	(nn)	n
DESTINATION	REGISTER	A	ED 57	ED 5F	7F	78	79	7A	7B	7C	7D	7E	0A	1A	DD 7E d	FD 7E d	3A n n	3E n
		B			47	40	41	42	43	44	45	46			DD 46 d	FD 46 d		06 n
		C			4F	48	49	4A	4B	4C	4D	4E			DD 4E d	FD 4E d		0E n
		D			57	50	51	52	53	54	55	56			DD 56 d	FD 56 d		16 n
		E			5F	58	59	5A	5B	5C	5D	5E			DD 5E d	FD 5E d		1E n
		H			67	60	61	62	63	64	65	66			DD 66 d	FD 66 d		26 n
		L			6F	68	69	6A	6B	6C	6D	6E			DD 6E d	FD 6E d		2E n
	REG. INDIRECT	(HL)			77	70	71	72	73	74	75							36 n
		(BC)			02													
		(DE)			12													
	INDEXED	(IX + d)			DD 77 d	DD 70 d	DD 71 d	DD 72 d	DD 73 d	DD 74 d	DD 75 d							DD 38 d n
		(IY + d)			FD 77 d	FD 70 d	FD 71 d	FD 72 d	FD 73 d	FD 74 d	FD 75 d							FD 36 d n
	EXT. ADDR.	(nn)			32 n n													
	IMPLIED	I			ED 47													
		R			ED 4F													

SOURCE

Figure 3.12   Instruction using extended addressing

only 8. Therefore, with the instructions under the extended addressing mode, although one address is specified, the data from that address is loaded into the *low byte* of the register pair and the data from the **next address** is loaded into the *high byte* of the register pair. For example:

LD HL,(2345H).

This instruction loads the L register with the data from address 2345H, then loads the H register with the data from address 2346H. The instructions are highlighted with thick lines in *Figure 3.14* (page 53).

Register Exchanges

There is a small group of instructions which allow data to be transferred between registers with a single byte instruction (*Figure 3.15*). The one that is used most frequently is probably EX DE,HL (hex code EB), which exchanges the contents of the DE and HL register pairs.

The Z80 has an alternate set of registers, in addition to the main register set, which duplicate the main registers in every respect. However, there are only two instructions that allow data in the alternate register set to be manipulated, and they are the two found in the exchanges group, with hex codes 08 and D9 respectively.

To exchange the contents of the AF registers (accumulator and flags) with the alternate accumulator and flag register, the mnemonic EX AF,AF' is used. This has the effect of simply transferring the main accumulator and flag register into the alternate register set, but note that the alternate register set contents are also transferred into the main registers at the same time.

The other register pairs in the main register set, BC, DE and HL can be transferred into the alternate registers BC', DE' and HL' with a single instruction whose mnemonic is EXX, and hex code D9. This has exactly the same effect as the instruction swapping the accumulator and flags, apart from the fact that it operates on all three register pairs at the same time. The main use for this instruction is during subroutines, which are discussed in detail in Chapter 5.

Input and Output Instructions

The Z80 has a very powerful set of input and output instructions, which allow data to be transferred from any register to an output port or read into any register from an input port. In addition, blocks of data can be read directly into or out of memory. However, by far the simplest way of input and output to the Z80 is to use the accumulator either as the source of the data or as the destination for the data. This method of data input/output uses immediate addressing, where the **port address** is written in the byte immediately after the op-code.

For example, to load the accumulator with data from port 80H, the instruction:

IN A,(80H) would be used, with hex code DB 80.

Similarly, to output data to port 81H, the instruction:

OUT (81H),A   Hex code D3 81

would be used. These two instructions are highlighted in *Figure 3.16*.

The use of the other instructions on these charts is not strictly necessary at this stage, since all the input/output required by most programs can be accommodated by the instructions shown.

		IMPLIED ADDRESSING				
		AF'	BC',DE', &HL'	HL	IX	IY
I M P L I E D	AF	08				
	BC, DE & HL		D9			
	DE			EB		
REG. INDIR	(SP)			E3	DD E3	FD E3

Figure 3.15   Register exchange instructions

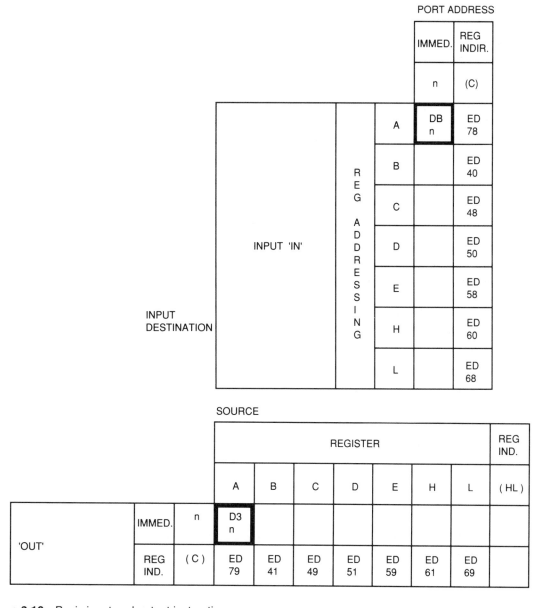

Figure 3.16 Basic input and output instructions

3.7 BINARY ARITHMETIC

Before examining some of the arithmetic instructions that the microprocessor is capable of executing, it is worth looking at the concepts of binary arithmetic so that it will be possible to predict what the computer is supposed to be doing when it performs arithmetic functions. The functions that the microprocessor can perform are extremely limited, and amount to only binary addition and binary subtraction. However, with these two basic operations it is possible to generate all the other arithmetic functions that are necessary. For example, multiplication can be carried out by repeated addition, and division can be carried out by repeated subtraction (although these may not be the best methods of performing these functions). Once the basic mathematical

A	B	Carry IN	SUM	Carry OUT
0	0	0	0	0
0	1	0	1	0
1	0	0	1	0
1	1	0	0	1
0	0	1	1	0
0	1	1	0	1
1	0	1	0	1
1	1	1	1	1

Figure 3.24 Full adder truth table

calculation illustrated in *Figure 3.23*. The carry input to the least significant bits can be either 0, or it can be made to come from the **carry flag**. The carry generated by each pair of bits is passed on to the next most significant bit within the register. If a carry is generated by the most significant bits being added together, then this carry is stored in the **carry flag**.

By providing a facility whereby the carry input can be added to a calculation, the microprocessor opens the way for multibyte arithmetic so that the calculations are not purely limited to 8 bit numbers.

Binary Subtraction

Binary numbers can be subtracted from one another in much the same way as decimal numbers can. The only problem arises when the number to be subtracted is larger than the number we wish to subtract it from. This then involves 'borrowing' from the next highest digit. The process must be familiar to most people using decimal numbers, but it is illustrated in binary in *Figure 3.25* where the number borrowed is a 2, from the next digit to the left. This is because the binary number system uses the base 2 whereas the decimal system uses the base 10.

A truth table for the subtraction of 2 binary bits can be produced, and this is shown in *Figure 3.26*. It is very similar to the half adder truth table

```
          2 2   2 2 2
  0 1 0 0 1 0 0 0        +72
  0 0 1 1 0 1 1 1 –      +55 –
     1 1   1 1 1    Borrow
  ─────────────────          ─────
  0 0 0 1 0 0 0 1    Result  +17
```

Figure 3.25 Binary subtraction

generated previously, but to be of any practical use it must be extended to accommodate the borrow bits from any previous calculation.

In practice it is not necessary to worry about how subtraction is carried out, since an alternative method exists which involves only addition. This is a process known as 'subtraction by complement addition'.

A	B	Difference	Borrow
0	0	0	0
0	1	1	1
1	0	1	0
1	1	0	1

Figure 3.26 Binary subtraction truth table

Compare the two calculations in *Figure 3.27*. The left-hand calculation shows a standard binary subtraction which uses 'borrows' from one pair of bits to the next most significant. In the right-hand calculation, instead of subtracting the second line of the calculation, its **complement** (all bits inverted) has been added to the first line of the calculation and an extra one has been added to it as well. Notice that the results turn out the same.

The only real difference is that the complement addition method results in an extra carry being generated in this case which is ignored.

Either method of subtraction can be used when doing manual calculations, just choose the one most convenient for you. As long as signed numbers are being used then the calculations will always result in the correct sign of the number in the answer.

Figure 3.27 Binary subtraction methods

3.8 THE FLAG REGISTER

The flag register is a vital part of the microprocessor, since the bits in the register hold information on the result which has just taken place in the arithmetic logic unit (ALU). The results stored by the bits of the flag register can be tested and used as the basis of decisions to be made in the computer, and it is this process that gives the computer its powerful decision-making capabilities.

Different microprocessors have different sets of flags. Each flag is simply an individual bit, which can either be **set** to a logical 1 or **reset** to a 0, depending upon the last arithmetic or logical operation which has taken place in the ALU. These bits can be examined by other microprocessor instructions and these can be made to perform in different ways depending upon the flag register contents. Although the flag register contains 8 bits, not all of them need to be used, and *Figure 3.28* shows the Z80 flag register.

Although the six flags shown are part of the flag register, they each act independently, so their position in the register is not particularly important. Of the flags, the most useful are the carry flag and the zero flag.

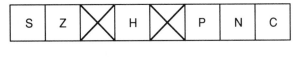

```
S   Z  [X]  H  [X]  P   N   C
```

S - Sign   P - Parity
Z - Zero   N - Add/subtract
H - Half carry   C - Carry

Figure 3.28 Z80 flag register

The Sign Flag – S

Signed two's complement representation for binary numbers uses the most significant bit (bit 7) of a binary number to indicate whether a number is positive or negative.

> If bit 7 = 0   the number is positive
>
> If bit 7 = 1   the number is negative

When the arithmetic logic unit of the CPU carries out a mathematical operation it copies bit 7 of the result into bit 7 of the flag register. Consequently, if signed number representation is being used the sign bit will indicate if the result of the last arithmetic operation produced a positive or negative number.

> **Sign** flag = 0   positive result
>
> **Sign** flag = 1   negative result

This facility allows the programmer to test the **sign** flag for a 1 or a 0 and then jump to a new portion of the program accordingly. For example, if the answer to a calculation was positive the program could carry out one task, but if the answer was negative the program may carry out a different task.

The Zero Flag – Z

There are many situations where the programmer must know if the answer to a calculation is 0. This facility is provided by the **zero** flag. Whenever the ALU produces an answer of 0 0 0 0 0 0 0 0 in binary, the **zero** flag will be set to a logic 1. If the answer is anything else then the **zero** flag is cleared to a logic 0.

The programmer can test for a zero result by using one of the conditional jump instructions, 'Jump on Zero' or 'Jump on Non Zero'.

The Half Carry Flag – H

The half carry flag is used in binary coded decimal arithmetic, which will be examined later. It is set to a logic 1 if a carry is generated between bit 3 and bit 4 of the calculation during an arithmetic operation. For example, consider the following addition:

$$
\begin{array}{r}
0\,0\,0\,0\,1\,0\,0\,0 \\
0\,0\,0\,0\,1\,0\,0\,0 \ + \\
1 \qquad \text{Carry} \\
\hline
0\,0\,0\,1\,0\,0\,0\,0
\end{array}
$$

The carry was generated between bit 3 and bit 4 of the calculation and this would result in the half carry flag being set.

Unlike other flags in the flag register the half carry flag cannot be used by a jump instruction, but it is used by the **decimal adjust accumulator** instruction (DAA).

The Parity Flag – P

Parity is the name given to a technique that enables the microprocessor chip, or any other electronic circuit to check the number of logical 1s contained in a byte of data. The technique is widely used for detecting errors when data is being transmitted from one digital system to another. If the parity flag is set to a logical 1 there is an even number of logical 1's in the ALU result. If the parity flag is cleared to a logic 0, then there is an odd number of logic 1s in the ALU result. The parity flag operates only after the execution of a logical instruction.

The Overflow Flag – V

The overflow flag shares the same bit as the **parity** flag but it only operates after the execution of an **arithmetic** instruction. In technical terms, it is the 'exclusive OR between the carry into bit 7 and the carry out of bit 7'. In practice, it indicates a sign error when two's complement numbers are being used, e.g. the number resulting in the accumulator has exceeded +127 or is less than −128.

The Add/Subtract Flag – N

The add/subtract flag is used to indicate the nature of the previous instruction performed by the ALU. If the previous operation was a subtraction or decrement instruction then the N flag = logic 1. Alternatively, for an addition or increment instruction N flag = 0.

This flag is not strictly the same as the other flags since it depends upon the instruction type and not the result, but it is used by the **decimal adjust accumulator** instruction to make the necessary corrections for decimal arithmetic.

The Carry Flag – C

The carry flag is probably the most frequently used of all the flags by the programmer. It is used to indicate a carry being generated from bit 7 of the ALU result. Note that it will indicate a carry for an addition or increment instruction and also indicate a borrow for a subtract, or decrement instruction.

In addition, when the contents of registers are shifted either to the left or the right, the bit that is displaced from the register can be made to go into the carry flag. This is frequently used as a method of testing the contents of selected bits in registers.

Flags in the flag register are only affected by certain microprocessor operations. In particular, they are affected by any **arithmetic or logic operation**, but there are some exceptions to this rule. The precise detail of how any flag is affected by an instruction is given at the beginning of the Z80 instruction set, which should be referred to for precise information.

3.9 ARITHMETIC INSTRUCTIONS

Because the microprocessor is a general-purpose device, it is not particularly specialised in terms of its mathematical capability. In fact, its mathematical functions are severely limited, and consist of only addition and subtraction, together with increment (add 1) and decrement (subtract 1). Some of the most modern types of microprocessor also include multiply and divide functions, but these are not included in the Z80 microprocessor instruction set. If the programmer wishes to perform these functions then it is necessary to write a short program which includes only addition instructions.

The vital thing to remember with all arithmetic operations, using 8 bits, is that *one of the operands is always in the accumulator*. This means that whatever is added is added to the accumulator, and the result is stored in the accumulator. The same applies to subtraction operations, in which everything is subtracted from the value in the accumulator.

Addition Instructions

There are two addition instructions.

The first adds two numbers together in binary, and the second adds the two numbers plus the contents of the carry flag. For these instructions a variety of addressing modes can be used, and they are shown highlighted in *Figure 3.29*.

The correct mnemonic for each instruction is:

ADD A,s   Add
ADC A,s   Add with carry

	SOURCE										
	REGISTER ADDRESSING							REG. INDIR.	INDEXED		IMMED.
	A	B	C	D	E	H	L	(HL)	(IX+d)	(IY+d)	n
'ADD'	87	80	81	82	83	84	85	86	DD 86 d	FD 86 d	C6 n
ADD W CARRY 'ADC'	8F	88	89	8A	8B	8C	8D	8E	DD 8E d	FD 8E d	CE n
SUBTRACT 'SUB'	97	90	91	92	93	94	95	96	DD 96 d	FD 96 d	D6 n
SUB W CARRY 'SBC'	9F	98	99	9A	9B	9C	9D	9E	DD 9E d	FD 9E d	DE n
'AND'	A7	A0	A1	A2	A3	A4	A5	A6	DD A6 d	FD A6 d	E6 n
'XOR'	AF	A8	A9	AA	AB	AC	AD	AE	DD AE d	FD AE d	EE n
'OR'	B7	B0	B1	B2	B3	B4	B5	B6	DD B6 d	FD B6 d	F6 n
COMPARE 'CP'	BF	B8	B9	BA	BB	BC	BD	BE	DD BE d	FD BE d	FE n
INCREMENT 'INC'	3C	04	0C	14	1C	24	2C	34	DD 34 d	FD 34 d	
DECREMENT 'DEC'	3D	05	0D	15	1D	25	2D	35	DD 35 d	FD 35 d	

Figure 3.29   Add instructions

Where s is either the register to be added, or the appropriate mnemonic for register indirect, indexed or immediate addressing modes. In other words, the letter or symbols taken from the second line of the chart as with the **load** instructions. There is no destination register indicated since it is implied that the destination of all the instructions on the chart is the accumulator. The only exceptions to this are the increment and decrement instructions.

For example, to add 16 hex to the value in the accumulator the instruction would be:

ADD A,16H   Hex code C6 16

For example, to add the contents of the H register to the accumulator including the carry bit, the instruction would be:

ADC A,H   Hex code 8C

In each case, the SOURCE of the data to be added to the accumulator contents is found on the chart as the heading of one of the columns. The instruction required is found as the label of one of the

SOURCE

	REGISTER ADDRESSING							REG. INDIR.	INDEXED		IMMED
	A	B	C	D	E	H	L	(HL)	(IX+d)	(IY+d)	n
'ADD'	87	80	81	82	83	84	85	86	DD 86 d	FD 86 d	C6 n
ADD W CARRY 'ADC'	8F	88	89	8A	8B	8C	8D	8E	DD 8E d	FD 8E d	CE n
SUBTRACT 'SUB'	97	90	91	92	93	94	95	96	DD 96 d	FD 96 d	D6 n
SUB W CARRY 'SBC'	9F	98	99	9A	9B	9C	9D	9E	DD 9E d	FD 9E d	DE n
'AND'	A7	A0	A1	A2	A3	A4	A5	A6	DD A6 d	FD A6 d	E6 n
'XOR'	AF	A8	A9	AA	AB	AC	AD	AE	DD AE d	FD AE d	EE n
'OR'	B7	B0	B1	B2	B3	B4	B5	B6	DD B6 d	FD B6 d	F6 n
COMPARE 'CP'	BF	B8	B9	BA	BB	BC	BD	BE	DD BE d	FD BE d	FE n
INCREMENT 'INC'	3C	04	0C	14	1C	24	2C	34	DD 34 d	FD 34 d	
DECREMENT 'DEC'	3D	05	0D	15	1D	25	2D	35	DD 35 d	FD 35 d	

Figure 3.30   Subtraction instructions

rows. The correct machine code instruction is then given where the row and column intersect.

Subtract Instructions

The subtraction instructions are almost identical with those that perform addition. The subtraction process can be carried out in simple binary or the carry flag can also be subtracted from the accumulator as well as the other number subtracted. The instructions are highlighted in *Figure 3.30* (opposite) with heavier lines. The correct mnemonic for each instruction is:

SUB s   subtract in binary
SBC A,s   subtract with carry (borrow)

For example, to subtract the number 33 hex from the accumulator the instruction would be:

SUB 33H

Notice that the instruction mnemonic is different

SOURCE

	REGISTER ADDRESSING							REG. INDIR.	INDEXED		IMMED
	A	B	C	D	E	H	L	(HL)	(IX+d)	(IY+d)	n
'ADD'	87	80	81	82	83	84	85	86	DD 86 d	FD 86 d	C6 n
ADD W CARRY 'ADC'	8F	88	89	8A	8B	8C	8D	8E	DD 8E d	FD 8E d	CE n
SUBTRACT 'SUB'	97	90	91	92	93	94	95	96	DD 96 d	FD 96 d	D6 n
SUB W CARRY 'SBC'	9F	98	99	9A	9B	9C	9D	9E	DD 9E d	FD 9E d	DE n
'AND'	A7	A0	A1	A2	A3	A4	A5	A6	DD A6 d	FD A6 d	E6 n
'XOR'	AF	A8	A9	AA	AB	AC	AD	AE	DD AE d	FD AE d	EE n
'OR'	B7	B0	B1	B2	B3	B4	B5	B6	DD B6 d	FD B6 d	F6 n
COMPARE 'CP'	BF	B8	B9	BA	BB	BC	BD	BE	DD BE d	FD BE d	FE n
INCREMENT 'INC'	3C	04	0C	14	1C	24	2C	34	DD 34 d	FD 34 d	
DECREMENT 'DEC'	3D	05	0D	15	1D	25	2D	35	DD 35 d	FD 35 d	

Figure 3.31   Increment and decrement instructions

3.11 ASCII CODE – American Standard Code for Information Interchange

It was stated earlier that the way in which the 8 bits in each byte of data are used is entirely up to the programmer, and this is true.

So far, it has been seen how the 8 bits in a byte of data can be used to represent:

- A microprocessor instruction.
- A binary number.
- A two's complement number.
- A pair of BCD digits.

There is one other major use that must be considered and that is to use the byte to represent alpha-numeric characters. This means that the byte contains a code that represents the letters, numbers and punctuation normally found on a computer keyboard. The standard code used to

represent these characters is known as ASCII code.

ASCII code uses 7 bits to represent the majority of alpha-numeric characters, although some systems use 8 bits so that some graphic or special characters can be included. The universal 7-bit code is shown in *Figure 3.34*. Notice from the table that all the NUMBERS start with a 3, so $30 = 0$, $31 = 1$, $32 = 2$, etc.

The **capital** letters start with $41 = A$ and then continue to $5A = Z$. Similarly, **lower case** letters go from $61 = a$ to $7A = z$. The first two columns are known as the control characters, and they are used to control the operation of printers and other peripherals. For example, code $0A = LF$ or line feed and code $0D = CR$ or carriage return which control the print operations on a VDU or printer.

Whenever data representing messages, text or other language based information must be stored or manipulated in a computer system, that data is

Bit numbers											0	0	0	0	1	1	1	1
											0	0	1	1	0	0	1	1
											0	1	0	1	0	1	0	1
b_7	b_6	b_5	b_4	b_3	b_2	b_1		hex 1 / hex 0			0	1	2	3	4	5	6	7
			0	0	0	0				0	NUL	DLE	SP	0	@	P	`	p
			0	0	0	1				1	SOH	DC1	!	1	A	Q	a	q
			0	0	1	0				2	STX	DC2	"	2	B	R	b	r
			0	0	1	1				3	ETX	DC3	£	3	C	S	c	s
			0	1	0	0				4	EOT	DC4	$	4	D	T	d	t
			0	1	0	1				5	ENQ	NAK	%	5	E	U	e	u
			0	1	1	0				6	ACK	SYN	&	6	F	V	f	v
			0	1	1	1				7	BEL	ETB	'	7	G	W	g	x
			1	0	0	0				8	BS	CAN	(	8	H	X	h	y
			1	0	0	1				9	HT	EM	)	9	I	Y	i	z
			1	0	1	0				A	LF	SUB	*	:	J	Z	j	{
			1	0	1	1				B	VT	ESC	+	;	K	[	k	|
			1	1	0	0				C	FF	FS	,	<	L	\	l	}
			1	1	0	1				D	CR	GS	-	=	M	]	m	~
			1	1	1	0				E	SO	RS	.	>	N	^	n	DEL
			1	1	1	1				F	SI	US	/	?	O	_	o	

Figure 3.34   ASCII code table

invariably stored as ASCII data. Therefore, the computer programmer must be aware of the codes used and must be able to manipulate ASCII data with the same ease as other data stored by the system.

3.12 LOGIC INSTRUCTIONS

Many of the programs run by a microcomputer system do not deal directly with numerical values. The 8 bits in a byte of data may each represent the ON or OFF state of a switch, a relay, a motor or a valve, for example. Therefore, instructions are required that allow the programmer to manipulate individual bits in a byte of data and these are known as the **logic instructions**.

The logic instructions provided by a microprocessor are basically the same as those provided with logic gates in a combinational logic circuit, and therefore they include **AND**, **OR**, **NOT** and **EXCLUSIVE OR**. In addition there is a **COMPARE** instruction and the facility to **SHIFT** bits in a byte of data to the left or to the right. Each of these instructions has a special function and use within a program, so they will be considered separately. They are all shown highlighted (thicker lines) in *Figure 3.35*, which also shows the correct mnemonic in the left-hand column. Notice with the logic instructions the reference to the accumulator A, is omitted in the complete mnemonic.

SOURCE

	REGISTER ADDRESSING							REG. INDIR.	INDEXED		IMMED
	A	B	C	D	E	H	L	(HL)	(IX+d)	(IY+d)	n
'ADD'	87	80	81	82	83	84	85	86	DD 86 d	FD 86 d	C6 n
ADD W CARRY 'ADC'	8F	88	89	8A	8B	8C	8D	8E	DD 8E d	FD 8E d	CE n
SUBTRACT 'SUB'	97	90	91	92	93	94	95	96	DD 96 d	FD 96 d	D6 n
SUB W CARRY 'SBC'	9F	98	99	9A	9B	9C	9D	9E	DD 9E d	FD 9E d	DE n
'AND'	A7	A0	A1	A2	A3	A4	A5	A6	DD A6 d	FD A6 d	E6 n
'XOR'	AF	A8	A9	AA	AB	AC	AD	AE	DD AE d	FD AE d	EE n
'OR'	B7	B0	B1	B2	B3	B4	B5	B6	DD B6 d	FD B6 d	F6 n
COMPARE 'CP'	BF	B8	B9	BA	BB	BC	BD	BE	DD BE d	FD BE d	FE n
INCREMENT 'INC'	3C	04	0C	14	1C	24	2C	34	DD 34 d	FD 34 d	
DECREMENT 'DEC'	3D	05	0D	15	1D	25	2D	35	DD 35 d	FD 35 d	

Figure 3.35 Logic instructions

The Logic AND Instruction

An AND gate in a combinational logic circuit works on the principle that 'an output is a logic 1 only if input A **AND** input B are both at logic 1'. This is shown in *Figure 3.36*. In the microprocessor, the **AND** operation takes place between each bit of the accumulator, and a corresponding bit in another byte of data. The other data may come from a number of sources, either:

(a) a byte of data from an internal register,
(b) an immediate byte of data from the program,
(c) a byte of data from a memory location.

A	B	Output
0	0	0
0	1	0
1	0	0
1	1	1

Figure 3.36 **AND** gate truth table

The result is always placed back in the accumulator. The three data sources listed above use three different types of addressing:

(a) Register addressing is used for data from a register.
    For example, AND B – AND the contents of the accumulator with the data in the B register.
(b) Immediate addressing is used for data from the program.
    For example, **AND OFH** – **AND** the accumulator contents with the number 0F hex.
(c) Register indirect addressing is used for data from memory.
    For example **AND (HL)** – **AND** the accumulator contents with data from the memory address pointed to by the HL register pair.

Bit Masking

One of the primary uses of the **AND** instruction is in a technique known as **bit masking**. In this technique the data ANDed with the accumulator contents is used to 'force bits to 0' in the accumulator data. To see how this works, consider the example below.

This could be the effect of the instruction AND F0H:

Initial Accumulator Contents (A)	1 0 1 0 1 1 1 1
AND with F0 (MASK DATA)	1 1 1 1 0 0 0 0
Result in Accumulator A0	1 0 1 0 0 0 0 0

When the **AND** operation takes place, notice that for every logic 0 in the second line of the calculation, there is a logic 0 in the result. In other words, *the logic 0s have forced logic 0s into the answer*. However, where a logic 1 appears in the second line of the calculation, the original data from the accumulator is unchanged.

In other words – a logic 0 in the mask will force a logic 0 into the accumulator.

A logic 1 in the mask will not change the data in the accumulator.

Notice also that the logic operation affects the flags in the flag register and in particular, the zero flag. This means that if the accumulator result is 0 0 0 0 0 0 0 0, the zero flag will be set to 1. Bit masking is a very widely used facility. Consider the program below:

```
IN A,(80H)
AND 7FH
OUT (81H),A
```

The three lines of program input data from port 80 hex, then whatever the input data, bit 7 is forced to be a logic 0, before the data is output to port 81 hex. This is often used when transmitting ASCII data for example, to ensure that only the valid ASCII codes are sent down a transmission line.

The Logic OR Instruction

The logic **OR** instruction is based on the OR gate truth table, in the same way that the **AND** operation is based on the AND gate truth table. This is shown in *Figure 3.37*. Here the output is a logic 1 if either

A OR B is a logic 1

A	B	Output
0	0	0
0	1	1
1	0	1
1	1	1

Figure 3.37 **OR** gate truth table

As with the **AND** instruction, the microprocessor performs the **OR** operation bit by bit with data in the accumulator and data from some other source.

For example, the instruction **OR** F0H may have been executed:

Initial Accumulator Contents (A)	0 0 1 1 1 0 0 0
OR F0H (MASK DATA)	1 1 1 1 0 0 0 0
Result in Accumulator F8	1 1 1 1 1 0 0 0

Notice that wherever a logic 1 appears in the mask data logic 1 also appears in the result, but where logic 0 appears in the mask data then the original accumulator data remains unchanged. This operation can be summarised as follows:

A logic 1 in the mask will force a logic 1 into the accumulator.
A logic 0 in the mask will not change the data in the accumulator.

Frequently the **OR** instruction is used after the **AND** instruction when particular data must be manipulated. First the **AND** instruction is used to force bits to logic 0, then the **OR** instruction is used to restore bits to a 1.

The EXCLUSIVE OR (XOR) Instruction

The **EXCLUSIVE OR** function is another widely used logic device, which works as follows:

'The output is a logic 1 if either input A **OR** input B are logic 1 but not when input A **AND** input B are both logic 1.'

The truth table for this function is shown in *Figure 3.38*.

The **EXCLUSIVE OR** function has an interesting effect when it is used in data manipulation. As before, examine the operation of a typical **EXCLUSIVE OR** instruction as shown below.

A	B	Output
0	0	0
0	1	1
1	0	1
1	1	0

Figure 3.38 **XOR** gate truth table

Again, suppose the instruction XOR F0H has been executed:

Initial Accumulator Contents (A)	0 1 1 0 0 1 0 1
XOR F0H	1 1 1 1 0 0 0 0
Result in Accumulator 95H	1 0 0 1 0 1 0 1

When a logic 1 appears in the mask data the number in the result is the previous data **inverted**. However, when a logic 0 appears in the mask data the original data in the accumulator is **unchanged**. Therefore, the **XOR** instruction can be used to selectively invert individual bits in the accumulator. Its operation can be summarised as follows:

A logic 1 in the mask will invert a bit in the accumulator.
A logic 0 in the mask will not change the data in the accumulator.

The NOT Instruction

All of the logic instructions can be followed by a **NOT** instruction if their inverse is required. The instruction to perform this operation is known as '**complement the accumulator**'. Its mnemonic and op-code are:

CPL   Hex code 2F

It can be found in the general-purpose AF operations instruction chart. The instruction is only rarely required however, since there are other ways to perform the same operation such as using the **EXCLUSIVE OR** instruction, for example:

XOR   FFH

will perform the same operation, although it requires an extra byte in memory.

The COMPARE Instruction

All the instructions considered so far will manipulate the data contained in the accumulator, but will also leave a modified result in the accumulator. Occasionally it is useful to be able to COMPARE the data in the accumulator with any other byte of data, and simply to set the **flags**, leaving the accumulator contents unchanged. This is the function of the **compare** instruction.

The **compare** instruction carries out a simple binary subtraction between the contents of the accumulator and another byte of data either from a register, memory address, or from an immediate instruction in the program. Note that the accumulator contains the top line of the calculation. However, the result of the subtraction is **not stored** in the accumulator as you would expect, but the **flags** are modified in the usual way for a subtraction. Consequently, if the byte of data being compared is the same as the contents of the accumulator the zero flag will be set.

The **zero** and **carry** flags indicate the relative magnitude of the numbers being compared. For example, if the number in A is compared with the number N, the **flags** will be affected as shown in *Table 3.1* opposite.

By executing a **compare** instruction and then examining the flags using a **conditional jump** it is possible to take different paths through a program depending upon the number compared with the accumulator contents.

Source and Destination

TYPE OF ROTATE OR SHIFT		A	B	C	D	E	H	L	HL	(IX + d)	(IY + d)
	'RLC'	CB 07	CB 00	CB 01	CB 02	CB 03	CB 04	CB 05	CB O6	DD CB d 06	FD CB d 06
	'RRC'	CB 0F	CB 08	CB 09	CB 0A	CB 0B	CB 0C	CB 0D	CB OE	DD CB d 0E	FD CB d 0E
	'RL'	CB 17	CB 10	CB 11	CB 12	CB 13	CB 14	CB 15	CB 16	DD CB d 16	FD CB d 16
	'RR'	CB 1F	CB 18	CB 19	CB 1A	CB 1B	CB 1C	CB 1D	CB 1E	DD CB d 1E	FD CB d 1E
	'SLA'	CB 27	CB 20	CB 21	CB 22	CB 23	CB 24	CB 25	CB 26	DD CB d 26	FD CB d 26
	'SRA'	CB 2F	CB 28	CB 29	CB 2A	CB 2B	CB 2C	CB 2D	CB 2E	DD CB d 2E	FD CB d 2E
	'SRL'	CB 3F	CB 38	CB 39	CB 3A	CB 3B	CB 3C	CB 3D	CB 3E	DD CB d 3E	FD CB d 3E
	'RLD'								ED 6F		
	'RRD'								ED 67		

	A
RLCA	07
RRCA	0F
RLA	17
RRA	1F

Figure 3.39 (a) Rotate and shift instructions

Table 3.1 Effect of flags when comparing two numbers A and N

Condition	Zero flag	Carry flag
A = N	1	0
A > N	0	0
A < N	0	1

The ROTATE and SHIFT Instructions

The microprocessor can be made to simulate the operation of a **shift register** by executing one of its many **rotate** or **shift** instructions. These simply move the contents of the selected register one place to the left or to the right. The variety of instructions arise because of the number of possible operations that can take place with the bits that 'fall out' of the end of the register. The carry flag plays a very important role in all of these instructions since generally it holds the bit that is lost from the register, or it holds the bit that is fed into the register. All of the **rotate** and **shift** instructions are illustrated in *Figure 3.39* (pages 70–1), together with the pictorial representation of

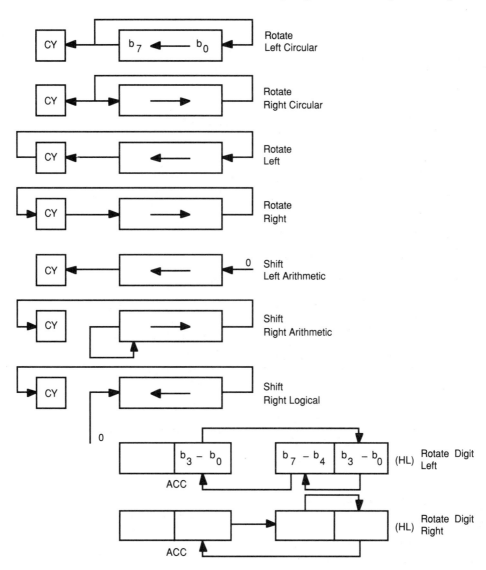

Figure 3.39 (b) Rotate and shift instructions

what they achieve. Although there are very many instructions, the four most commonly used are those single byte instructions that operate directly on the accumulator, shown to the right of main instruction block.

There are three basic instruction types:

- The **rotate circular** instructions rotate the contents of the register one bit in either direction, and the displaced bit is input to the other end of the register. In addition it is copied into the carry flag.
- The **rotate left** and **rotate right** instructions perform a similar function, but the carry flag becomes the ninth bit of the register and is included in the **rotate** operation.
- The **shift** instructions simply shift the register contents right or left and the data displayed is stored in the carry flag.

Figure 3.40 shows the operation of the RLCA instruction 'Rotate Left Circular the Accumulator'. Imagine that the accumulator originally contains 0F hex, and the carry flag is 0. The figure shows the result of five RLCA instructions in a program.

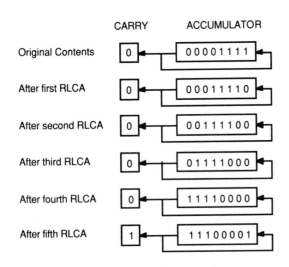

Figure 3.40   Operation of RLCA instructions

3.13   TEST AND BRANCH INSTRUCTIONS

The instructions in this group are sometimes referred as the 'Transfer of Control Instructions', or simply the 'Jump Instructions'. This is because they are instructions that allow the programmer to jump from one part of the program to another, either unconditionally, or depending upon the state of one of the flags in the flag register. Without jump instructions, all computer programs would simply execute one instruction after another from beginning to end and there would be no possibility of changing, and therefore no possibility of decision making. It is the process of using a conditional jump instruction to test a flag that enables the processor to make a decision based upon a previous arithmetic or logical result.

The most common are the unconditional jump instruction and the instruction that jumps depending upon the state of zero flag. These are just two of quite a large group, which are shown in *Figure 3.41*.

Unconditional Jumps

All the instructions in this group contain an op-code followed by a two byte address which follows the normal convention of having the low-order byte first. For example, to force the program to jump unconditionally to address 2010H, the instruction would be:

JP 2010H C3 10 20

When this instruction is encountered in a program, the processor jumps to the new address without checking the contents of any flags. *The new address may be either before or after the current address*, which makes it possible to create a continuous loop program by Jumping to an earlier address.

By jumping to an address after the current address, it is possible to omit a certain block of memory. This memory may be used to contain data for a program or some other routine which is not to be used during the current program.

Conditional Jumps

The **conditional jump** instructions perform an entirely different function. They allow one of two routes through a program to be selected depending upon whether the chosen condition (the state of individual flags) is met or not.

CONDITION

			UN-COND.	CARRY	NON CARRY	ZERO	NON ZERO	PARITY EVEN	PARITY ODD	SIGN NEG.	SIGN POS.	REG B = O
JUMP 'JP'	IMMED. EXT.	nn	C3 n n	DA n n	D2 n n	CA n n	C2 n n	EA n n	E2 n n	FA n n	F2 n n	
JUMP 'JR'	RELATIVE	PC + e	18 e - 2	38 e - 2	30 e - 2	28 e - 2	20 e - 2					
JUMP 'JP'	REG. INDIR.	(HL)	E9									
JUMP 'JP'		(IX)	DD E9									
JUMP 'JP'		(IY)	FD E9									
DECREMENT B, JUMP IF NON ZERO 'DJNZ'	RELATIVE	PC + e										10 e - 2

Figure 3.41 The jump instructions

If the condition is met   the processor jumps
If the condition is NOT met   the program continues with the next instruction.

Consider the problem of where a processor must jump to address 1900 hex if the input data from port 80H is 55 hex, but must continue with the normal program for any other input data.

A suitable program is shown below:

IN A,(80H)
CP 55H
JP Z,1900H
LD B,30H   This is executed if A is NOT 55H
etc.

When the input data is **compared** with the number 55H, it will set the zero flag if it is **equal**. Therefore this can be tested in the JP Z, instruction so that when the zero flag is SET the program jumps to address 1900H. For all other input values the instruction LD B,30H is executed and the program continues as normal.

Imagine that the program must jump to address 1900H if the input data is less than the number 55H. Almost the same program can be used, but this time instead of jumping if the zero flag is 1, the jump can be made if the carry flag is 1, since it will only be set if the accumulator data is less than the number 55H. Thus the program would be modified as follows:

IN A,(80H)
CP 55H
JP C,1900H
LD B,30H
etc.

There are a whole range of conditions that allow the programmer flexibility in testing the flags for either a logic 1 (**set**) state or a logic 0 (**reset**) state. The correct mnemonic for each of these conditions is given below:

JP Z	Zero flag set	(Zero)
JP NZ	Zero flag NOT set	(Non Zero)
JP C	Carry flag set	(Carry)
JP NC	Carry flag NOT set	(No Carry)
JP PO	Parity ODD	(Flag 0)
JP PE	Parity EVEN	(Flag 1)
JP P	Sign Flag positive	(Flag 0)
JP M	Sign negative (minus)	(Flag 1)

Summary

The main points covered in this chapter are:

- The microprocessor executes those instructions defined by the binary codes in its instruction set.

- Microprocessor instructions can be grouped as:
  (a) Data transfer instructions.
  (b) Data manipulation instructions.
  (c) Flow of control instructions.

- Assembly language is used to program microprocessors because it offers high-speed operation coupled with low memory requirements.

- Machine code programs can be produced from an assembly language program either by hand translation using charts or by the use of an assembler program.

- The addressing mode is the means by which the microprocessor finds the data or address it is to operate upon.

- Z80 addressing modes include register, immediate, indirect, extended address, implied, indexed and relative addressing.

- Data transfer instructions are used to move data around a microprocessor system.

- Arithmetic and logic instructions are used to manipulate data or perform calculations.

- Jump instructions force a processor to jump to another part of the program to continue execution.

- The bytes of data in a computer may represent 8-bit numbers, signed binary numbers, BCD numbers, ASCII data or machine instructions.

Questions

3.1 Briefly describe what is meant by microprocessor 'instruction set'.

3.2 List the three main classes of instruction found in all microprocessor sets.

3.3 Why is assembly language programming often preferred to a high level programming language?

3.4 What is meant by a 'symbolic address'?

3.5 List some of the functions of a **monitor** program.

3.6 What facilities must be present in a microprocessor system if it is to be used for assembly language programming?

3.7 Write down the mnemonic and hex code for the following instructions:

(a) Load register C from register H.

(b) Load register A from register L.

(c) Load register L from register A.

3.8 Write down the hex code for the following mnemonics:

(a) LD H,37H.

(b) LD E,01H.

(c) LD L,76H.

3.9 Write down the mnemonic which would represent the following hex codes:

(a) 3E 59.

(b) 1E 1E.

(c) 06 B1

3.10 Assume that the HL register pair is loaded with the address 1B00H. Write down the mnemonic and hexadecimal instructions to:

(a) Load register D from address 1B00H.

(b) Load register C from address 1B00H.

(c) Load the contents of the accumulator into address 1B00H.

(d) Load address 1B00H with a number 67H.

3.11 Briefly describe the effect of the following instructions:

```
LD B,18H
LD C,00H
LD H,19H
LD L,00H
LD A,(BC)
LD (HL),A
```

3.12 Write down the two instructions that will perform the following task:

Load the accumulator from address 1910H and then place the data in address 1920H.

3.13 Write down the mnemonics for two sequences of instructions that will load the C register with data from address 1905H.

3.14 Write down the correct mnemonic for the hex code below.

11 77 20

3.15 Write down the mnemonic and hex code for the instruction that loads the HL register pair with 1800H.

3.16 Describe the effect of the instruction LD (1234H),BC and write down its hexadecimal code.

3.17 Write down the correct mnemonic and hex code for the following instructions:

(a) Output data from the accumulator to port 63H.

(b) Exchange the HL and DE register pair contents.

(c) Load the B register with data 53H.

(d) Load the BC register pair with a number 6417H.

(e) Load the HL register pair with the data from memory address 6417H and 6418H.

(f) Put the contents of the accumulator into memory address 6417H

3.18 Write down the following decimal numbers in signed two's complement form:

(a) +105

(b) −18

(c) +86

3.19 Perform the following calculations in binary:

(a) 77 + 52

(b) 25 + (−13)

(c) 81 − 70

(d) −12 − (−33)

3.20 Write down the correct mnemonics and hex code for the following instructions:

(a) Add the contents of the E register to the accumulator.

(b) Add with carry the number 81H to the accumulator.

(c) Add the contents of the memory address pointed to by the HL register pair to the accumulator.

3.21 Write down a series of instructions that will form the following tasks, using the correct mnemonics only:

(a) Add 20 hex to the contents of the B register.

(b) Add the value from input port 80 hex to the contents of the H register.

3.22 Write down the instruction mnemonic and hex code that would be used to:

(a) Subtract with carry the contents of the memory address pointed to by the HL register pair from the accumulator.

(b) Subtract the accumulator contents from itself.

3.23 Write down the series of instruction mnemonics which would be required to subtract 5 from the contents of the L register.

3.24 Write down the instruction mnemonic and hex code for the command to:

(a) Increment the accumulator.

(b) Decrement the value in the memory address pointed to by the HL register pair.

3.25 Write down the hex code and the instruction mnemonic for the following:

(a) Subtract the contents of the DE register pair from the HL register pair.

(b) Decrement the DE register pair.

(c) Add the contents of the index register IX to itself.

(d) Decrement the stack pointer.

3.26 Write down the correct mnemonics for the instructions that will:

(a) **AND** the accumulator with the data in the D register.

(b) **OR** the data in the memory address pointed to by the HL register with the accumulator.

(c) Invert all the accumulator bits except bit 0.

3.27 If the E register contains 77 hex, the D register contains F0 hex and the accumulator contains AA hex, what will be the result in the accumulator when the following program is executed:

```
OR D
AND E
```

3.28 What is the effect of the instruction XOR A?

3.29 Write the mnemonics for a program that would perform as follows:

Data is input from port 80 hex, then bits 0 to 3 are output to port 81 hex, but in bit positions 4 to 7. All other bits should be 0.

3.30 Why is it necessary to use the ADC (add with carry) instruction instead of the **ADD** instruction when performing multi-byte arithmetic operations?

3.31 Briefly explain how a computer that normally calculates with binary numbers can be programmed to deal with decimal numbers.

3.32 Which logical instructions would need to be used to perform the following operations on the data in the accumulator?

(a) Force bits 2, 3 and 4 to logic 0.

(b) Force bits 1, 6 and 7 to logic 1.

(c) Invert bits 3 and 5.

3.33 What data will be output from port 81H if the input data from port 80H is 37
hex when the following program is run?

```
IN A,(80H)
AND 1FH
OR 80H
XOR 61H
OUT (81H),A
HALT
```

3.34 What is meant by the term 'bit masking'?

Writing and running computer programs

4.1 WRITING PROGRAMS

Serious Z80 assembly language programmers should be very familiar with the range of instructions which the Z80 microprocessor can execute. They have been covered in some detail because they are the foundation of much more complex programs. Most of the instructions used so far have been explained in the context of a short program, but it is only when larger programs are examined that it is possible to develop a greater understanding of programming techniques.

Programming skills and techniques must be developed so that it is possible to write programs when only a simple explanation of the problem has been supplied. In many cases, the people who

request specific computer programs will have no idea about how those programs could be written, they will only know what the required end result is, and this would usually be stated in plain English. It is therefore vital to be able to take a plain English description of a problem, and break it down into a number of steps that will allow a satisfactory solution to be produced.

The sequence of steps that together form the solution to the problem, is known as the **ALGORITHM**.

The algorithm describes the method to be used to solve a problem. In a number of cases, there may be a variety of different algorithms that may be capable of solving the problem, although one of the solutions will usually be preferable in terms of its speed of execution, or simplicity in program design.

The use of two different types of algorithms to produce the same result may be illustrated by considering how to multiply two 8-bit binary numbers together.

One method uses the 'repeated addition' for simple multiplication. This involves adding a number to itself by the number of times it must be multiplied. The flow chart for this operation is shown in *Figure 4.1*.

The second method is known as the 'shift and add method' and its flow chart is shown in *Figure 4.2*. It is not the intention here to examine how this program works, but simply to illustrate the fact that a completely different algorithm can be used to create the same result. The important thing to

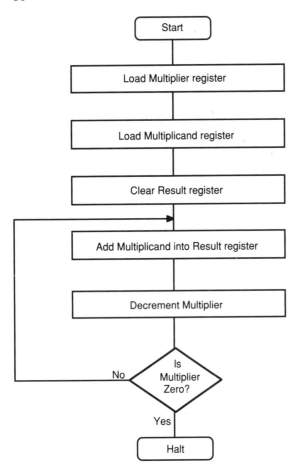

Figure 4.1   Repeated addition flow chart

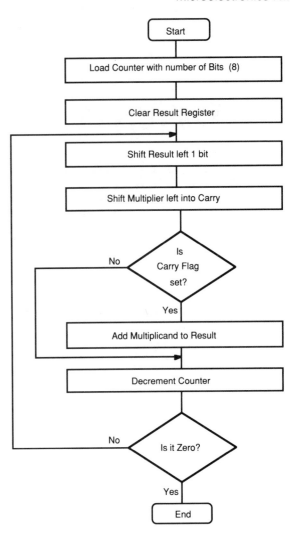

Figure 4.2   Shift and add method flow chart

notice is that the method is totally different, but the results should be identical.

Although the second flow chart looks much longer, when the machine code is produced from it it turns out to be only an extra 2 bytes. However, when the program is executed, the execution time for the first method is considerably longer than the second when the numbers are greater than 8.

Therefore the algorithm and the look of the flow chart can be very deceptive, and it is particularly important to choose an algorithm that is suitable for the problem in hand. It must be chosen for maximum speed if this is an important criterion, or the shortness of the code if the memory utilisation must be low, etc.

Writing down an algorithm to solve a given problem involves breaking the problem down into a series of steps. Generally, this is best achieved

by producing a flow chart. By examining the previous two figures it can be seen that the steps in the program are written in plain English although they are also related to the type of instructions that may be used in the final program.

The subject of flow charts is particularly important and will be examined in some detail.

4.2 FLOW CHARTS

Drawing a flow chart is usually the first step in producing a computer program, because it allows programmers to organise their thoughts and produce a logical organisation for the program. Flow charts may be devised at different levels of complexity. For example if very general statements are placed in the flow-chart boxes, then the overall function of each part of a program can be devised. Then as more detail is required, the complexity of each flow-chart box can be broken down into a number of more specific flow charts at a machine code level. In fact, flow charts are by no means restricted to producing programs in machine code, since they are almost universally applicable to any computer language. The main virtue of a flow chart is that it allows the programmer to see the logical flow of ideas, and this allows possible errors caused by poor program design to be identified at an early stage.

Flow charts have been used for many years in the documentation of computer programs. As such, various standards exist and specific symbols are given on flow charts for particular functions. Some of these are shown in *Figure 4.3* although this is by no means an exhaustive list. They represent the main flow chart symbols that are in common use.

In most of the flow charts that have been used so far the oblong boxes are used to hold the majority of instructions. Those that describe 'test and branch' instructions, or conditional statements, use the diamond-shaped boxes which have

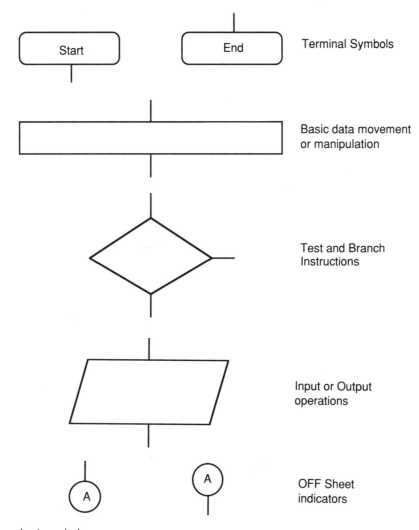

Figure 4.3 Flow chart symbols

two alternative exits. One of the exits is used when the condition being tested is true, and the other exit is used when the condition is not true. However, this is often clarified with a 'Yes' or 'No' beside the branches. The input or output operations box is rarely used since both input and output are often seen as basic data movement instructions using an oblong box.

The flow-chart boxes hold a description of the operation to be carried out at a particular point in the program, and are interconnected by uni-directional lines which indicate the flow of the program.

Flow-charting Techniques

Although it is possible to write programs in many different ways, there are a number of functions that tend to be used repeatedly in all manner of different applications. For example, the task of performing an operation a given number of times is a very common device, as are the operations to select one of two options or one of three options in a program. Since these are in such common use their operation may be described by a very simple flow chart, which then may be translated into the appropriate code in programs.

Loop Counting

Many programs include a function which has to be performed repetitively. The basic flow chart for a simple loop counter is shown in *Figure 4.4*. This is the most common technique for performing an operation any number of times.

The number of times the operation is to be repeated is first loaded into a register known as the counter. The instructions which constitute the repeated operation then follow and these may be one or two, or may be very extensive. Following the repeated operation, the counter is decremented and if it has not reached zero, the loop is repeated. Only when the counter reaches zero does the program continue.

There are only two things to be careful of in this example. The first is that the contents of the counter must not be affected by the repeated operation. If it is then the whole of the counting

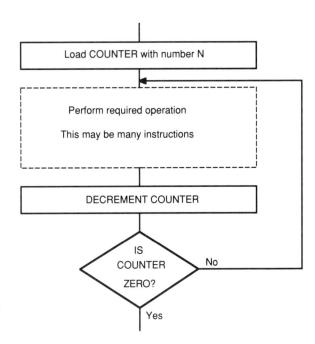

Figure 4.4 Simple loop counter flow chart

will fail. Secondly, the technique only works for a counter value that is one or more. An initial value of zero in the counter will cause the loop to be repeated a large number of times, depending upon how many bits are used for the counter register. This can clearly be a disadvantage especially if the counter contents are not known precisely by the programmer, which may be the case if the data is input from a port or some other place.

The flow chart shown in *Figure 4.5* (opposite) addresses this problem, since it tests the value for zero before it performs the required operation. This allows a value of zero to skip the whole of the loop, while any other value in the counter will cause the program to be repeated that number of times.

Most counting systems use the first method described, since the count value is generally known. A special case of this program exists when a simple delay is required in a program, since the required task is simply to waste some time, which can be fulfilled by adding some 'no operation' instructions, or simply by having no instructions at all in the loop.

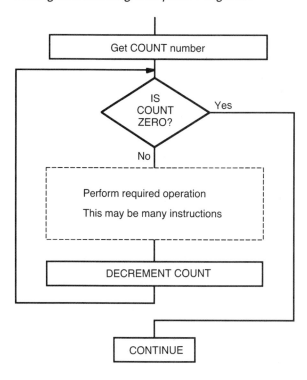

Figure 4.5   COUNTER with initial zero test

Creating a Time Delay

It is quite often necessary to make the microprocessor carry out certain operations for a fixed length of time. This is achieved by sending the starting command for the operation and then making the microprocessor delay in a counting loop.

Consider the program below:

```
        LD B,00
LOOP:   DEC B
        JP NZ,LOOP
```

This represents a simple counting loop which will cause a delay of about 2 ms. The exact time delay can be calculated by multiplying the number of loops (256 in this case) by the time per loop. The time per loop depends upon the number of T-states required by each instruction. In the example above:

```
LD B,00      –   7 T-states
DEC B        –   4 T-states
JP NZ,LOOP   –   10 T-states
```

Thus the total program requires $7 + 256\,(4 + 10)$ = 3591 T-states. If the clock rate is 2.0 MHz each T-state takes 500 ns.

Total time for the program is 3591×500 ns = 1.8 ms.

A longer time delay can be created by counting down using a register pair as the counter instead of a single register. The only drawback with this method is that the decrement instruction for a register pair **does not** set the flags. In particular another means must be found to discover when the count has reached zero. This technique is shown in the program below:

```
        LD BC,000
LOOP:   DEC BC
        LD A,B
        OR C
        JP NZ,LOOP
```

The two instructions

```
LD A,B
OR C
```

are designed to perform the logical **OR** function between registers B and C, the counter registers. If a logic 1 exists anywhere in the register pair then the zero flag will not be set. However, when the count reaches 0 0 0 0 hex, the zero flag will be set.

The instructions in the previous program take the following number of T-states:

```
LD BC,0000   10
DEC BC        6
LD A,B        4
OR C          4
JP NZ,LOOP   10
```

With a 2.0 MHz clock, the time delay created would be about 0.8 seconds.

Making Decisions

Another very basic operation which is required in many computer programs is the ability to make simple decisions. Generally, there are two alternatives, to perform Task A or Task B depending upon whether certain conditions are met. The logical flow chart is shown in *Figure 4.6* (page 84).

This flow chart shows that input data is tested to see whether it meets certain conditions. If the

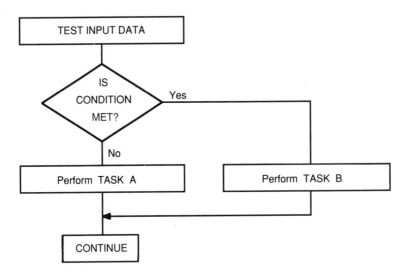

Figure 4.6   Choosing ONE of TWO OPTIONS

conditions are met then Task B is performed but if they are not met Task A is performed. The flow chart shows these two tests alongside one another, but in practice they would have to occupy different locations in the computer memory. Therefore in reality, the flow chart could be re-

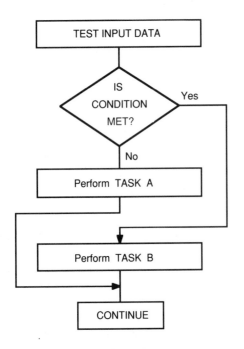

Figure 4.7   Practical choice of TWO OPTIONS

drawn as shown in *Figure 4.7*. Here the tasks are shown to follow one another in the memory, but by virtue of the way the lines are drawn, you can see that a jump is required after Task A has been performed 'around' Task B. If this jump is not included in the program, then when the conditions are not met both Task A and Task B will be executed, which is clearly not the required operation.

Choosing one of three options in a program illustrates this point further. Two tests are required and care must be taken to ensure that the program does not proceed to execute all the options, since they must be coded following one another in the computer memory. This is illustrated in *Figure 4.8*.

Notice from the flow chart that two conditions are checked at the beginning of the procedure, but only one test of the input data is made. This is because generally the single test can set flags in the flag register which can be tested in the two conditional operations which follow. Certain programs, however, may required an additional test of the input data between the two conditional instructions. Once again, jumps after Task A and Task B ensure that the subsequent tasks are not undertaken in addition to the one that is required.

Programs of this type are frequently encountered when the compare instruction is used as the test for the input data. The **compare** instruction

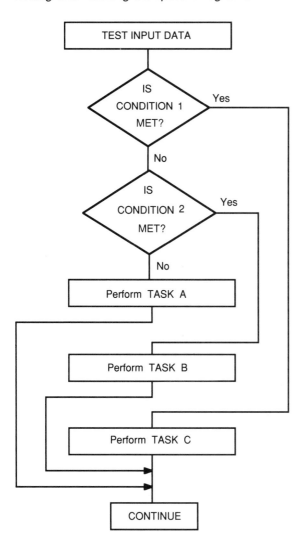

Figure 4.8   Choosing ONE of THREE OPTIONS

sets both the **Carry** and the **Zero** flags, so these may both be tested by subsequent conditional jump instructions, without the need for an additional test of the input data.

The preceding examples are simply to illustrate the way in which certain operations may be performed in computer programs. They illustrate the most common operations that will be required, but there are many variations upon them. Here, they only serve to illustrate common programming practice.

Frequently there are many ways to perform the same task by using slightly different programs,

and there seem to be advantages and disadvantages with the method selected. It is only by practice and experience that the best methods are often obtained. The following case study will serve to illustrate this point.

4.3 CASE STUDY 1: GENERATING AN ANALOGUE WAVEFORM FROM A COMPUTER

At the beginning of the book it was explained how the computer can deal only with digital information. Any analogue values that must be input have to be converted to digital form before they can be processed by the computer. Similarly, an output waveform in analogue form must be generated from its digital equivalent in a device known as a 'digital-to-analogue converter'.

In an 8-bit microcomputer system, the 8 bits of data can represent 256 different values. Therefore, if the output analogue voltage can vary between 0 and 2.55 volts, each increment in the digital number would represent a change in the analogue voltage of 10 millivolts. This automatically limits the accuracy with which any analogue voltage can be represented to within + or −5 mVs. The only way this can be improved is by increasing the number of bits used to represent the analogue voltage or by reducing the range of voltages represented. *Figure 4.9* (page 86) shows how the digital numbers in the computer are used to represent the different analogue voltages.

Each bit of the binary number has a particular weight and the further to the left the bit is in the number, the greater its significance. Bit 7, for example, represents a change of half the maximum voltage which the digital-to-analogue converter can represent whereas Bit 0 represents only 1/256th of this value. The weight of each of the bits which make up the binary number are shown in *Figure 4.10* (page 86).

In the digital-to-analogue (D to A) converter, each bit then contributes a voltage towards the final output which is generated by the device. The individual contributions are added together in a 'summing amplifier', which accurately combines the individual voltages (*Figure 4.11*, page 86). Whenever the digital input signal to the D to A

Digital Equivalent	Analogue Voltage	Proportion of Maximum
0 0 0 0 0 0 0 0	0.00	0
0 0 0 0 0 0 0 1	0.01	V_{max} /256
0 0 0 0 0 0 1 0	0.02	2 V_{max} /256
0 0 0 0 0 0 1 1	0.03	3 V_{max} /256
⋮	⋮	⋮
1 1 1 1 1 1 1 0	2.54	254 V_{max} /256
1 1 1 1 1 1 1 1	2.55	255 V_{max} /256

Figure 4.9   Digital and analogue equivalents

BIT	7	6	5	4	3	2	1	0
Analogue voltage produced	$\dfrac{V_{max}}{2}$	$\dfrac{V_{max}}{4}$	$\dfrac{V_{max}}{8}$	$\dfrac{V_{max}}{16}$	$\dfrac{V_{max}}{32}$	$\dfrac{V_{max}}{64}$	$\dfrac{V_{max}}{128}$	$\dfrac{V_{max}}{256}$

Figure 4.10   Weight of each bit

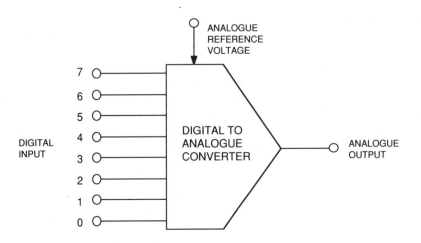

Figure 4.11   Digital-to-analogue converter

converter changes, there is a short time delay known as the 'settling time' before the output responds. This, however, is very short for most D to A converters and typically only of the order of 100 nanoseconds. This effectively means that the analogue output follows the digital input almost instantaneously so that whatever digital number is placed on the input, the output will respond with its equivalent analogue voltage. Normally, a digital output is derived from the output port of a micro-processor, or from some type of counting circuit.

Waveform Generation

If the D to A converter responds very quickly with the correct analogue voltage corresponding to any digital number received by its inputs, then in order to generate an analogue waveform, all that is required are the appropriate digital inputs, corresponding to the wave shape. It is a relatively simple matter to get the computer to generate such a sequence of digital numbers, although there are certainly one or more ways in which this can be achieved. The best method to use depends largely upon the type of waveform required, and here two possible methods will be considered:

(a) Calculating each value before it is output.
(b) Outputting pre-calculated values from a table in memory.

These two alternatives are shown in *Figure 4.12*. Consider how easy it would be to use each of the techniques indicated by the flow charts A and B to generate typical waveforms that may be required in any systems. The two waveforms chosen for this case study are shown in *Figures 4.13* and *4.14* (page 88). Notice that each of these is made up of a series of steps, which represent the outputs from the D to A converter. The length of time the output stays static on one of those steps, is determined by how fast the computer outputs data to the D to A converter. If it outputs the data quickly then the time will be short and the frequency of the waveform can be very high. However, if it outputs the data very slowly the waveform shape remains the same but its frequency is very low.

Consider first of all how easy it would be for the flow chart shown in *Figure 4.12(a)* to be used to generate the waveform for the ramp (*Figure 4.13*).

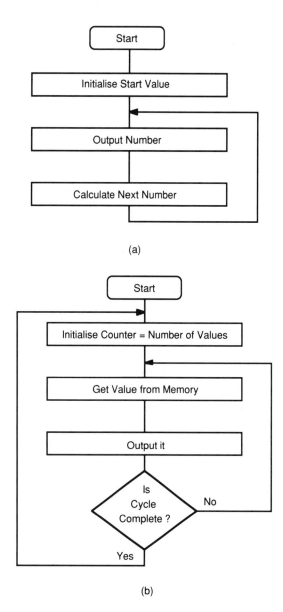

Figure 4.12 Generating an output waveform

Since the ramp simply represents an increase of 1 in the output number for each step, the problem of calculating the next number to be output is very simple. Therefore the program could be written in no more than about five or six lines. In fact the calculation would be so fast, that a delay would have to be introduced before each number was output in order to generate waveforms of different frequencies.

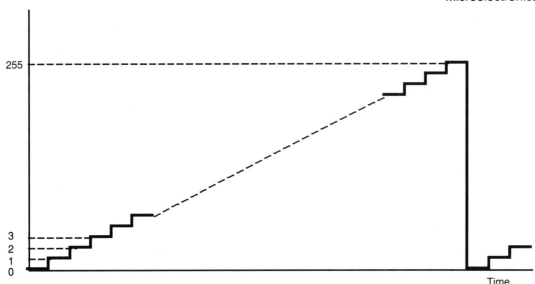

Figure 4.13   Ramp waveform (stepped)

However, using the flow chart in *Figure 4.12(a)* as the basis of a program to generate a sine waveform (*Figure 4.14*) is another story. From the waveform it can be seen that the steps are not equal, because the sine waveform has a very rounded shape.

Therefore the problem of calculating the next value becomes particularly difficult since the microprocessor has no way of calculating mathematical functions such as the sine of an angle, apart from the use of a power series method. This would

be a very time-consuming exercise in the microprocessor; and while it may be possible, it would take so long that the time between output pulses would be long and the highest frequency that could be generated would be relatively low.

Now consider the program that would be based on the flow chart in *Figure 4.12(b)*. Although this flow chart looks more complicated than that in *Figure 4.12(a)*, it can be used to generate waveforms of any shape with equal ease. In very simple cases, however, with a large number of values to output, the table of data values in memory becomes particularly long and therefore other methods may be more appropriate.

If this method were to be used to generate the ramp waveform, each output value starting from 0 and going up in steps of 1, would have to be stored in a table in memory. Clearly this would be 256 bytes long. The program would then fetch each byte from memory as required, output it, and then check to see if the cycle were complete. If it were not complete the process would be repeated until 256 values had been output, and the cycle could begin again. In comparison with the program of no more than six lines, to generate the values by calculation, this method seems particularly unwieldy and therefore not worthy of consideration. It may, however, be more useful if fewer

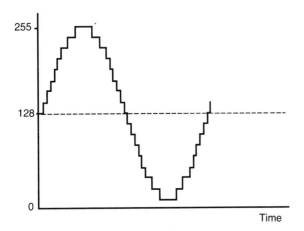

Figure 4.14   Sine waveform (stepped)

values were required to be output even if the waveform is simple.

However, if the method of storing the data in a table in memory were used to generate the sine waveform shown in *Figure 4.14*, then it becomes particularly useful. Before that can be achieved, the programmer must calculate the value required for each step. This can easily be done by means of a calculator or some mathematical tables and the appropriate values may be stored in a table in the computer memory. It is then a relatively simple matter for the program to retrieve each value as required, output it, and then delay for the required time. The time delay will allow waveforms of different frequency to be generated using the same basic data pattern.

So, with two different algorithms, waveforms can be generated by the computer, but each algorithm is more appropriate to certain types of waveform.

This, remember, is an illustration to emphasise the point that different algorithms may be used in computer programs that are both capable of realising a solution, but one of which is far better and more efficient than the other in certain circumstances.

4.4 CASE STUDY 2: READING INPUT DATA

Microprocessors are well known for their ability to input and manipulate data very quickly and then produce output results. In fact it is this ability that makes them so versatile in the field of data processing. It is a relatively simple matter to input a single byte of data, but the task of inputting a continuous stream of data bytes is more complex than may be first imagined.

The basic problem is that computers work so fast compared with human operators or mechanical input devices, that somehow their speed has to be controlled so that the data input can be properly synchronised. This could be done in a number of ways, but this case study investigates two possible algorithms which have a totally different technique of solving the same problem, each with their own advantages and disadvantages.

The first technique is probably more appropriate for information that will be provided to the computer on a regular basis. It involves introducing a fixed delay of a suitable length into the computer program so that its operation is slowed down sufficiently to match that of the input device.

The second technique involves setting up a signalling arrangement whereby the input device sends a signal to the microprocessor when new data is available. This arrangment is slightly more complicated than the first one but is widely used especially where the input of information may not happen regularly.

Introducing a Fixed Delay

The algorithm used to add a fixed delay to a program is best described by examining the flow chart shown in *Figure 4.15*.

Whenever any data is input to a microcomputer system, the program must know how many bytes of data to expect, or there must be some other way of terminating the input reading routine. In *Figure 4.15* this is achieved by initially loading a counter with the number of bytes that are expected. At the end of the flow chart the instructions to decrement the byte counter and then check to see if this is zero allow the processor to exit from the loop as soon as all the input data has been received. While in the loop, the data is read from the input and then the program delays for a fixed time. This time can be made variable depending upon the application and the consideration. For example, if the input is a human operator who is setting up data on switches then the delay may be 5 or 10 seconds, but if there is some mechanical device changing the data then the delay may be less than half a second, for example.

There are, however, a number of difficulties with this approach.

First, if the input data is not ready when the program wishes to read it in, there is no way to indicate to the program that it must continue to wait. Instead, it will simply input whatever data is available which may be incorrect. This also means that if any data is delayed or the program loses synchronism then incorrect results will occur.

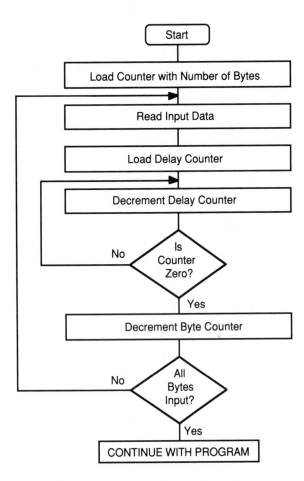

Figure 4.15 Inputting data at fixed intervals

There are, however, many applications where this type of approach may be ideal. For example, if the input data was from an analogue input transducer (a device that converts physical information such as ·temperature or velocity into electrical signals) this system could be used to sample the input data at regular intervals. Typically this method is used in devices that monitor physical properties in a process known as 'data logging'.

The main advantage of this system is its simplicity, and the fact that it requires no extra hardware connections between the computer and the device inputting the data. However, it is less than ideal in many circumstances because the computer must always be slowed down to a speed that is slower than the input so that no data is lost.

Waiting for an Input Signal

Some of the disadvantages of the previous method of inputting data may be overcome if the device wishing to send data to the computer also sends another signal to indicate when data is available. Sometimes this is known as a **'handshake'** signal, or a **'strobe pulse'**. Unfortunately this requires extra hardware, and one of the data bits may well be used for this purpose, restricting the input data to 7 rather than 8 bits. The technique is described using the flow chart in *Figure 4.16*.

The technique is very similar in many respects to that described previously, in that a counter is also required to count the number of data bytes that have been input. This will allow the microprocessor to terminate this part of the program correctly. However, instead of a delay, another loop has been included which inputs the data from the

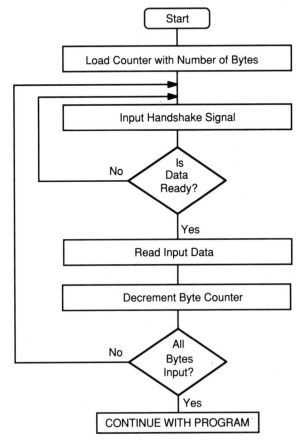

Figure 4.16 Waiting for a handshake signal

handshake signal and checks to see whether this indicates that data is ready.

In practice the handshake signal will be a single bit, so the test is for either a logic 1 or a logic 0 depending upon how the system hardware is set up. When the handshake signal is in the correct state, the program then continues to read the true input data, and the process is repeated until all the data bytes have been input. In more advanced systems, the processor would send a signal back to the data source to indicate that data had been received, and that it was ready for the next input byte.

The main advantage of this system is that the data input is completely synchronised, so that no time is wasted by the processor and data is input as soon as it is available. However, the simplicity of the technique masks a few potential problems, especially if the input handshake operation is implemented using a switch or other mechanical contact. This is because mechanical operations generally produce 'switch bounce' and extra delays have to be added to the system to accommodate this problem.

The case study has shown that there are a number of possible techniques to achieve the same objective, which are implemented using different algorithms. Clearly, some techniques have advantages over others in certain circumstances, and it is up to the programmer to decide which is the most appropriate algorithm for the given circumstances.

4.5 TRACE TABLES

When an appropriate algorithm has been chosen to solve a problem, its flow chart is normally translated into a computer program by writing it first in assembly language. This is then translated into machine code using the techniques described in Chapter 3.

However, every assembly language programmer soon discovers a problem – how to find out if the program is working correctly? The process of finding faults in a program and correcting them is generally known as **'debugging'** the program. In large programs, this process may take longer than actually writing the code itself, since if there are a large number of variables, each one needs to be tried to see whether the logic of the program works for all possible values.

While dealing with relatively simple programs, there are a number of techniques which can be used to check the correct operation of each one. The simplest of these is the idea of using a **trace table**. This is a means of checking to see whether the contents of the registers and memory addresses that have been used in the program actually contain the data that would be expected by the programmer. Producing a trace table is fundamental to the ideas of debugging a program, and therefore it is worthwhile investigating how to do it properly.

Generally, the main points of interest in a trace table are the contents of the CPU registers that are being used. It is important to discover whether they contain the expected information after each instruction has been executed. This is particularly important when the information in the register is used as part of the conditional jump instruction, or when an address is being used for the retrieval of information, for example. In addition, a trace table should contain information on the contents of any memory address which is changed as a result of the program operation, or the data present at a port during an input or output instruction.

Most computers provide a number of facilities which help in the production of trace tables. Many are accessible by using the appropriate key on the keyboard, or a simple MONITOR command.

For example, many single board computer systems would contain a small keyboard with **ADDRESS, GO, STEP, REGISTER** and **PC** keys which allow the programmer to examine the CPU and memory contents relatively easily.

Those useful for **trace table** operation are the **REGISTER, STEP** and **PC** keys in particular.

Some systems require a simple keyboard command to be entered and a program such as DDT (Dynamic Debugging Tool) or **DEBUG** to be used if trace tables are to be examined.

The [STEP] Key or Command

A [**STEP**] key or command executes a program in memory. However, it executes only one instruction at a time. When the [**STEP**] key is pressed, the instruction at the address currently in the program counter is executed, and the display then shows the next instruction address. It is possible to completely execute a program by continuing to press the [**STEP**] key, until a particular point in a program is reached. However, in the production of a trace table, it is more usual after each step, to examine the contents of the registers in the CPU or particular memory addresses.

Note that when a trace table is being produced, the contents of the address normally shown are the current program counter contents, but these represent the *next* instruction to be executed. However, when the registers are examined, they represent the result after the previous instruction has been executed. The program counter therefore always seems to be one instruction ahead of the contents of the registers.

The [REG] Key or Command

The [**REG**] key or command allows the programmer to examine the contents of any of the CPU internal registers. When producing trace tables it is normally only necessary to examine the contents of the registers, but it may also be possible to modify them.

The [PC] Key or Command

When the contents of a register or a memory address have been examined during a single-step operation, it is important to be able to return to the point in the program which has been reached. This is achieved with the [**PC**] key or command which stands for program counter. When this key is pressed or command entered, the system returns to the display showing the current program counter value, indicating that it is ready for another [**STEP**] key to be pressed so that program execution may continue.

Breakpoints

While it is possible to achieve a lot using only the [**STEP**] key or command, if a program is very long this can be a very laborious operation. It may often mean pressing the [**STEP**] key many hundreds of times. A better way of stepping a program at any desired point, is to use the **breakpoint**

facility. This is a feature that allows the programmer to set up an address at which the program will halt. For example, consider a program that included a long delay, and then the entry of a byte of data. If it were necessary to step through every instruction of the delay, before the instructions to input the data byte were able to be examined, this would take a very long time. However, by setting a **breakpoint** after the delay, the input instructions and their correct operation can easily be investigated.

In most small systems only one breakpoint can be set; and although this may seem to be a restriction, it is normally quite sufficient for program debugging. In more sophisticated systems, a number of breakpoints can usually be set, sometimes up to nine or ten, which will allow more sophisticated types of program to be checked for correct operation.

Consider the following example which illustrates the method of producing a trace table from a program without even testing it on a computer. When the actual program is run, the theoretical trace table can be checked against the actual values generated.

Example

Write down the trace table that you would expect to obtain if the program below was executed. Assume that all registers and memory locations contain 00 initially.

```
           ORG 1800H
START:     LD HL,1B00H
           LD (HL),25H
           LD B,(HL)
           LD DE,1B20H
           ADD A,50H
           INC B
           LD (DE),A
           LD (HL),B
           HALT
```

Each line of the program is treated separately and generates a set of values in the registers or memory addresses. In this case, the trace table should contain information on the Program Counter (PC), registers A, B, C, D, E, H, L and addresses 1B00H and 1B20H.

The first line ORG 1800H indicates that the PC will start at address 1800 hex.

The next instruction loads HL with 1B00 hex.

The next loads that address with 25 hex, and so on. When the trace table is complete it would look like that shown in *Table 4.1*.

Table 4.1 Trace table

PC	A	F	B	C	D	E	H	L	1B00	1B20
				Registers					| Memory	
1800	00	00	00	00	00	00	00	00	00	00
1803	00	00	00	00	00	00	1B	00	00	00
1805	00	00	00	00	00	00	1B	00	25	00
1806	00	00	25	00	00	00	1B	00	25	00
1807	25	00	25	00	00	00	1B	00	25	00
180A	25	00	25	00	1B	20	1B	00	25	00
180C	75	00	25	00	1B	20	1B	00	25	00
180D	75	00	26	00	1B	20	1B	00	25	00
180E	75	00	26	00	1B	20	1B	00	25	75
180F	75	00	26	00	1B	20	1B	00	26	75

4.6 THE MONITOR PROGRAM

In its most basic form, a microcomputer system consists of a CPU, a memory chip, and an input/output port. Unless the memory chip contains a program, a microcomputer in this form is not capable of any useful function. However, when a program is added, then the system may be made to perform a variety of tasks, according to the wishes of the programmer. This type of system is known as a **minimum system**.

Most single-board computers contain not only the minimum system components but also additional features such as a keyboard, a display, a tape input/output facility, a loudspeaker, etc. To make a system like this work properly it has to have a relatively sophisticated program resident in its ROM which is programmed with all the necessary functions. This program is known as a **monitor program**. Although there may be times when the computer system appears to be doing nothing, it is actually running the monitor program almost all the time. In particular when it is waiting for an entry on the keyboard, the monitor is always busy checking each key in turn, waiting for a key depression. Similarly, while any data is present on the display this is constantly being updated and the display is multiplexed (switched on one after another) which is controlled by the monitor program.

In larger disk based systems, the functions of a **monitor** program are often provided as part of the disk operating system, or the basic input/output system. Other features may form part of separate software packages.

The functions of a **monitor** program may be classified into two groups, the first being its basic operational facilities, and the second its machine code programming facilities.

The basic operational facilities of the monitor include the following:

(a) Reading keyboard entries.
(b) Outputting data to the display.
(c) Reading programs from casette tape.
(d) Writing data onto casette tape.
(e) Sounding the loudspeaker.

Each of these facilities is self-explanatory, and they are used whenever required by the monitor. In particular, the main loop which the monitor program executes continuously involves reading the keyboard and displaying the output data.

These facilities may be useful in their own right, but in many single board computers their main function is to assist in the entry of machine code programs in hexadecimal form. Therefore, there is another range of facilities built into the monitor program which assist the user to program efficiently using the machine code, and also give the ability to debug the programs when they are entered and executed. These facilities are as follows:

(a) Entry of hexadecimal data and translation into binary code which is stored in memory.
(b) Execution of a machine code program at a specific address.
(c) Examination and modification of data in memory.
(d) Examination and modification of registers in the CPU.
(e) Single-step facilities which execute one instruction at a time.
(f) Breakpoint facilities.

(g) Inserting and deleting data in memory.

(h) Moving blocks of data in memory.

(i) Generation of an interrupt signal.

The first six of these facilities are by far the most important. Almost all small microcomputer systems with a monitor program need to include each of these facilities if they are going to be useful for the machine code programmer. In larger systems, they may still be present, although in a slightly different form. For example, the facility to examine and modify memory which takes place one address at a time in a small system, may be replaced by complete screen display of memory addresses if a system with a video display unit is used.

4.7 BUS SIGNALS DURING PROGRAM OPERATION

The system buses in any microcomputer are almost always active. The only exceptions to this are immediately after the execution of a **halt** instruction, and during an extended **wait** operation. During the rest of its operating time, the microcomputer is busy interpreting instructions from its memory or from a user program. This activity can generally be monitored by examining each of the bus lines. One of the ways this activity can be seen in some systems is by examining lights which are driven by the signals on the bus lines.

Although lights on the **address**, **data** and **control** buses may flash slightly or seem to be dim they are actually responding to the data present on the buses, so that a bright light indicates the data is largely at the logic 1 level, whereas the light that is dim or out altogether indicates that the line is at the logic 0 level most of the time.

For example, when the monitor program is running, the high-order lights on the address bus may all be **off**. This is because the monitor program generally resides in the lower addresses in memory and therefore these are the lights that show the most activity. Similarly, the control bus shows that the activity is mainly memory oriented, i.e. monitor program mainly reads instruction from memory and executes them. However, some input/output activity takes place because data is read from the keyboard and sent to the display.

The best way to discover how the buses are affected by the programs that are running is to write a short program that only uses a certain range of addresses, and includes no input/output operations. The bus lights may then be examined to see if the correct activity is shown on each of them. However, care must be taken when interpreting the address lines A0 to A7, since these are not only used by the program being executed, but also contain the refresh addresses used by the Z80 to refresh dynamic memory devices.

Another simple way of observing bus activity is to use a logic probe. Although this device can only monitor the data at one point in the circuit at any moment, it can actually provide more information than an LED connected to a bus line. This is because most logic probes have a built-in pulse stretching facility and separate PULSE light so that even though a pulse may be very short, it will still flash the light of the logic probe. This allows the probe to capture information that would otherwise be missed simply by looking at the lights attached to a bus line.

The oscilloscope provides another method of examining the bus signals when programs are running, but again, most systems are limited to examining only one or two points at a time. With an oscilloscope the main problems arise because the waveforms are not generally repetitive enough to allow a stable trace to be achieved. If the program running is more than a few instructions long it almost always proves too difficult to synchronise the oscilloscope trace.

However, by synchronising an oscilloscope on another system signal such as an $\overline{IORQ}$ line or a chip select signal it is sometimes possible to use an oscilloscope trace to examine the waveforms in a system and to discover whether or not it is working correctly. In all systems, the greatest limitation of the oscilloscope is that it can only examine one or two circuit points at any moment.

A logic analyser is a device that was designed to allow many points in a microcomputer circuit to be examined simultaneously. Unfortunately they are expensive devices and when waveforms are examined they are often 'idealised' and not 'actual' waveforms. This is because the logic analyser samples the data on the bus lines and synthesises the waveforms from the samples.

The higher the sampling speed of the logic analyser, the better the resulting waveforms will be.

One major advantage of the logic analyser is that it is not limited to the actual waveforms in a system. It can often be used to interpret the data, and display it either in binary, hexadecimal or even assembly language format. This makes it a highly sophisticated and versatile tool for a microcomputer technician.

Summary

This chapter has covered the techniques of writing and running simple machine code programs, and some of the main points are as follows:

- Before efficient programs can be written, the instruction set of the microprocessor to be used must be thoroughly understood.

- The process of writing a program involves:
  (a) choosing a suitable algorithm.
  (b) producing a flow chart.
  (c) translating the flow chart to assembly language instructions.
  (d) translating the assembly language instructions into machine code.

- Flow charts are used to describe the basic structure of a program with universally understood boxes of different shapes containing the program operations.

- Standard algorithms exist for loop counting, choosing one of two or more options, etc.

- There is generally more than one way to write a program, although usually one of the alternative methods proves to be best.

- A trace table is a device used to note the contents of registers and memory addresses expected during the execution of a program.

- A monitor program provides facilities for keyboard entry, display output, and the entry of machine code programs generally in hexadecimal form. It also allows programs to be executed and corrected when necessary.

Questions

4.1 Briefly describe the function of an algorithm.

4.2 Under what circumstances would it be better for a computer to calculate the output values to produce an analogue waveform, rather than obtaining each value from a table?

4.3 List the **two** *essential* features of a monitor program which are necessary to allow machine code programs to be entered and executed.

4.4 Briefly describe what is meant by a **flow chart**.

4.5 Draw a **trace table** for the program below, showing all the registers used and the memory addresses 1A03 hex and 1803 hex. Assume the program starts at Address 1900H:

```
LD HL,1A03H
LD (HL),77H
LD D,18H
LD E,03H
LD C,(HL)
LD A,C
LD (DE),A
HALT
```

4.6 Why is a **breakpoint** used in software development?

FIVE

Subroutines and the stack

OBJECTIVES

When you have completed this chapter, you should be able to:

1. *Understand the function and application of subroutines, together with their main features.*
2. *Understand the term 'nested subroutines'.*
3. *Use the call and return group of instructions in the microprocessor instruction set.*
4. *Explain the operation of a last-in first-out (LIFO) stack and the stack pointer.*
5. *Understand the use of the stack to store:*
   (a) Subroutine return addresses.
   (b) Data from registers.
6. *Appreciate that subroutines can perform functions such as a:*
   (a) Timing delay.
   (b) Mathematical function.
   (c) Input/output routine.
7. *Understand a parameter passing technique to implement variations in subroutine functions.*

5.1 SUBROUTINES

In computer programs it is quite common to require a particular operation to be carried out a number of times at different points in the program. For example, a program may require a number of delays at different points, or a number of mathematical functions such as a multiplication or division. Using the techniques discussed so far, this would require that the particular code for the operation was entered into the computer program each time it was required. Clearly the code would be the same each time, so this would be very wasteful in terms of computer memory and also in programmer effort when entering the routine a number of times. In addition, each entry would have to be checked for its correct operation which could be very time consuming.

The way that all of these problems can be avoided is to use a technique known as a **subroutine**.

A subroutine is a special section of program which can be used as often as required by the main program. Generally the subroutines are entered into memory after the main program, and are then 'called' when required. This effectively puts the code of the subroutine into the main program at any point where it is required, but in the computer memory, the code for the subroutine exists only once.

This is illustrated in *Figures 5.1* and *5.2*. In *Figure 5.1* imagine that there is a delay which is required three times in the course of the program. The code for the delay would have to be included each time and the flow would proceed linearly throughout the program. This is very wasteful of memory and also each time the delay routine is entered its operation would have to be checked.

Compare this with the program structure in *Figure 5.2*, which although it looks more complicated is actually much simpler because the code for the delay exists only once. Each time the delay is required one instruction is simply entered into the main program. This is the **call** instruction

Instructions

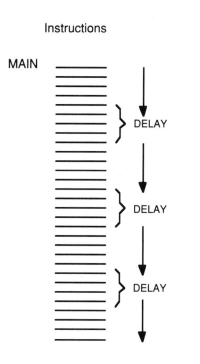

Figure 5.1 A 'linear' program without subroutines

which forces the flow of the program to the beginning of the subroutine. At the end of the subroutine the **return** instruction forces the program back to the instruction after the related call. The effect of the call is therefore to insert the complete delay routine in the main program. Mechanisms by which **calls** and **returns** work are discussed in more detail later.

Although a delay routine has been used as the example of a subroutine in the previous explanation, quite often the subroutine will perform a mathematical operation, an input, or an output function. In some cases, such as the mathematical routine, although the code stays the same the actual numbers used may be different. This means that some mechanism must exist to pass the numbers such as the **operands** to the subroutine, and then receive the results from the routine. This is a process known as **passing parameters**.

The use of subroutines in computer programs is generally to be encouraged since they have a number of advantages over other techniques. Some of these are listed overleaf:

Instructions

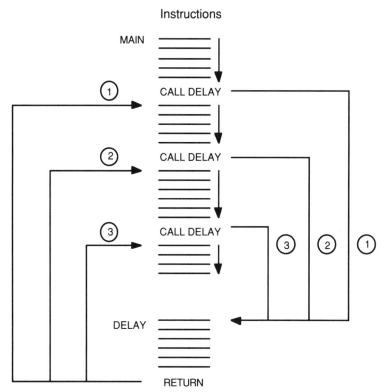

Figure 5.2 A program with a subroutine

(a) When subroutines are used the overall program uses less memory, since the subroutine exists only once but may be used many times. This is important particularly when programs become very long.

(b) Long programs can be sub-divided into subroutines, which will allow them to have a better program structure. This allows the programmer to concentrate on the main logic of the program without having to worry about the detail of each subroutine at the initial design stage. The function of the subroutine can be defined initially, then the detail can be worked out later. By writing programs in logical blocks, the programmer is forced to think of the program systematically, and this may even enable different programmers to work on different subroutines which will be later joined together as required by the main program.

(c) Since the code for each subroutine is only entered once, this can be individually tested using dummy data or a dummy main program. This improves the program's reliability considerably.

(d) Commonly used routines can be stored as libraries in larger computer systems that use disk storage. This relieves the programmer of the chore of re-writing simple routines time after time.

(e) There is only one slight disadvantage to the use of subroutines which is that programs including subroutines will take slightly longer to execute than the equivalent programs written without them. This is because there is a slight time penalty in the extra instructions to **call** and **return** from the subroutine which would not be present if a linear approach were used. However, in most cases, the advantages of using subroutines far outweigh this slight disadvantage and so the use of subroutines is normal practice in the vast majority of computer programs.

Nested Subroutines

Although *Figure 5.2* shows that subroutine is **called** from the main program, there is no restriction on where the subroutine is **called** from. It is therefore quite possible to **call** one subroutine from within another, and this is known as **nesting** the subroutine.

Imagine that a program was required to calculate trigonometrical functions of an angle, such as the sine, cosine and tangent. Part of this calculation would require a multiplication function. This means that the multiplication subroutine could be **called** from the routine which calculated the cosine for example. This is illustrated in *Figure 5.3*.

Notice that both the cosine routine and the multiply routine are actually subroutines of the main program. Each one has a **return** instruction at the end of it which forces the program to **return** to the point from which it was called. This means that when the multiply routine is called it

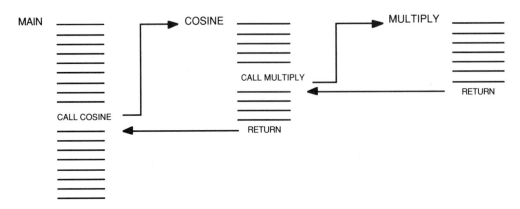

Figure 5.3 Example of a 'nested subroutine'

returns to the subroutine which called it, i.e. the cosine subroutine. Similarly, this **returns** to the main program when it is complete. There is no real limit to the number of subroutines which may be 'nested' within one another in most micro-processors. In the Z80 the only limitation is the amount of memory and this would restrict the number of 'nested' subroutines to many thousands. It is simply a function of the amount of available memory.

Note that in *Figure 5.3* there is no restriction on where the multiply routine needs to be called from. For example, it could be called directly from the main program, or if other subroutines exist for the calculation of sine and tangent of an angle, these may also use the multiply subroutine. The system is quite flexible and versatile.

5.2 THE CALL AND RETURN INSTRUCTIONS

Subroutines do not form part of the main computer program, but are usually located in memory somewhere after it. They are accessed when required using the **call** instruction, by **calling** the required start address of the subroutine. In assembly language programming, since labels can be used instead of addresses, the name of the subroutine can be generally used, which makes it easier to read the function of the main program. This point was illustrated in *Figure 5.3* where the subroutines were given names rather than addresses, although in practice the call instruction clearly requires an address to execute correctly.

The **call** and **return** instructions occupy a special instruction group which is shown in *Figure 5.4.*

The figure shows that the **call** instructions require an address following the op-code. This is the address of the beginning of the subroutine, and follows the normal format for addresses in the Z80 instructions where the low-order byte comes immediately after the op-code, followed by the high-order byte. The chart shows the unconditional **call** instruction, which is by far the most frequently used, together with all the conditional **call** instructions. These are the same conditions used for the jump instructions so that the programmer has complete flexibility about how subroutines are called.

Note that there is a major difference between a **CALL** *and a* **JUMP** *instruction. The call instruction*

CONDITION

			UN-COND.	CARRY	NON CARRY	ZERO	NON ZERO	PARITY EVEN	PARITY ODD	SIGN NEG.	SIGN POS.	REG B = O
'CALL'	IMMED. EXT.	nn	CD n n	DC n n	D4 n n	CC n n	C4 n n	EC n n	E4 n n	FC n n	F4 n n	
RETURN 'RET'	REGISTER INDIR.	(SP) (SP+1)	C9	D8	D0	C8	C0	E8	E0	F8	F0	
RETURN FROM INT 'RETI'	REG. INDIR.	(SP) (SP+1)	ED 4D									
RETURN FROM NON MASKABLE INT 'RETN'	REG. INDIR.	(SP) (SP+1)	ED 5D									

NOTE: CERTAIN FLAGS HAVE MORE THAN ONE PURPOSE. REFER TO Z80 CPU TECHNICAL MANUAL FOR DETAILS.

Figure 5.4 CALL and **RETURN** group

stores the address of the next instruction before it jumps to the beginning of the subroutine. The jump instruction simply jumps directly to the new address, and there is no mechanism for it to return automatically to any point in the program. The conditional instructions appear in pairs as before, with the condition of the carry, zero, parity and sign flags being capable of being tested for either a logic 1 or a logic 0 condition.

The **return** instructions follow a similar pattern with the unconditional **return** being by far the most commonly used. Notice that the return instructions have no address associated with them. This is because the address to which the program will return is automatically stored on the stack during the **call** instruction operation, which means that the **return** instruction simply activates the mechanism to retrieve this return address during its operation.

Consider the following examples which illustrate the use of the CALL instruction:

(a) CALL unconditionally address 1900H.
    Hex code   CD 00 19
(b) CALL address 6633H if the zero flag is set.
    Hex code   CC 33 66

The correct mnemonics for the call and return instructions are given below:

CALL 1234H	CALL unconditionally (address 1234H).
CALL NZ,1234H	CALL on non zero.
CALL Z,1234H	CALL on zero.
CALL NC,1234H	CALL on non carry.
CALL C, 1234H	CALL on carry.
CALL PO,1234H	CALL on parity odd.
CALL PE,1234H	CALL on parity even.
CALL P,1234H	CALL on sign positive.
CALL M,1234H	CALL on sign negative (minus).

The mnemonics for the **return** instructions are the same as those for the **call** except that the call is replaced by the mnemonic **RET**.

In the **call** and **return** group of instructions there are two instructions on their own at the bottom of the chart. These are **return** instructions which are used at the end of **interrupt** routines, but these will not be considered further here.

5.3 THE OPERATION OF THE STACK

It is interesting to consider how the microprocessor remembers the address to return to at the end of a subroutine. The answer is that it employs a device built into the microcomputer system known as the **stack**.

· The stack is so named because its operation resembles the operation of a stack of plates in a restaurant or cafeteria for example. When plates are required, they are taken off the top of the stack to be used, and this operation takes place until the stack is empty. However, when the stack begins to go down, extra plates may be added at the top. These would naturally be the first ones to be taken off by customers who require a plate, so the device operates such that a last plate to be put on the stack is the first one to be taken off. It is therefore known as a **last-in**, **first-out'** stack (*Figure 5.5*).

The Last-In, First-Out, LIFO Stack is a common device in computer systems, which is used to store information. There are other similar devices such as a **queue** which is a first-in, first-out device (FIFO), but this is beyond the scope of the current book.

In the microcomputer system, the stack is actually a series of memory locations which are used to store the addresses that are used for the return from a subroutine. With most microprocessors, the stack is placed in an area of random access memory well out of the way of other programs, typically at some of the highest memory addresses used in the system. However, the choice of where to place the stack depends upon the software system designer. In other microprocessors the stack may be implemented as part of the CPU, and therefore some storage locations are used within the CPU itself. However, this approach has a number of difficulties since the size of the stack may be very limited.

The mechanism by which the return addresses for subroutines are placed on the stack and returned at the appropriate time, requires a review of what happens during the machine cycles of a **call** instruction. There is a special register in the microprocessor known as the **stack pointer** and this is responsible for managing the whole of the operation of the Z80 stack.

FILLING THE STACK

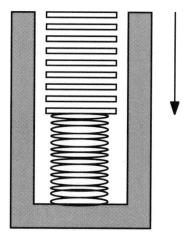

EMPTYING THE STACK   [LAST-IN FIRST-OUT]

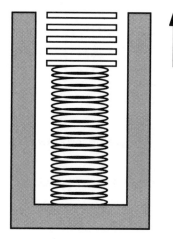

Figure 5.5 A typical STACK mechanism

Suppose the stack is to be located at address 1FFFH, in a typical microcomputer system. This would be achieved by initially loading the value 2000H into the stack pointer register, using one of the 16 bit load instructions. This value is held in the stack pointer until a call instruction is encountered in a program, and then the following mechanism takes place.

The program counter is used to hold the address of the instruction to be fetched from memory and then executed, but immediately each instruction is fetched, during T_2 of each machine cycle, the number in the program counter is automatically incremented. Suppose the instruction CALL 1234H is encountered in the program. The hex code for this would be CD 34 12. This is illustrated in *Figure 5.6*.

The program counter would initially have address 1830H in it, and the microprocessor would proceed to execute the instruction at this

address. During M1 the program counter is incremented to 1831H and after the **call** mnemonic is read, the first byte of the **call** address is also read, and the program counter is again incremented to address 1832. Here the final part of the address is read, and the program counter is incremented to address 1833H. At this point, the stack pointer begins its operation, with the value 1833H held in the program counter.

First the stack pointer is decremented to address 1FFFH, and the **high byte** of the program counter value is written into the address pointed to by the stack pointer. The stack pointer is decremented once again, to address 1FFE0H, then the **low byte** of the value in the program counter is written into that address. This is illustrated in *Figure 5.7* (page 102), which shows the situation before the call instruction is encountered and that just after it has been executed.

Notice that the contents of the program counter are placed on the stack with the high byte first, and also that it appears that the subroutine start address has not been stored. This, however, is not the case, since the start address of the subroutine, address 1234H, has been stored in the temporary address registers W and Z in the Z80. As soon as the program counter contents have been placed on the stack, the W and Z register contents are placed on the address bus to begin the next part of the program at that address. *In other words the*

Address	Hex Code	Mnemonic
1830	CD	CALL 1234H
1831	34	
1832	12	
1833	47	LD B,A
1834	–	

Figure 5.6 CALL instruction operation

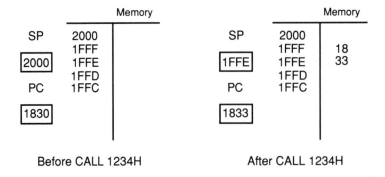

Before CALL 1234H   After CALL 1234H

Figure 5.7   Stack operation during a CALL

program jumps to address 1234H having previously saved the correct return address on the stack. During the execution of the first instruction of the subroutine the program counter is automatically updated with the new address, i.e. address 1235H, during T_2 of the next machine cycle.

The operation of the stack may be further clarified by examining what takes place during 'nested subroutines'. *Figure 5.8* shows part of a program where three **call** instructions follow immediately after one another. The program at address 1820H **calls** the subroutine at address 1830H, which immediately calls the subroutine at address 1850H, and this immediately calls the subroutine at 1880H. After the first call instruction, if the stack pointer was originally 2000H, it would be decremented to address 1FFE as explained previously, and the address 1823H would be placed on the stack. As soon as the next call instruction is encountered the stack pointer is decremented again and the return address for the second call is placed on the stack immediately after the first return address. The same takes place again when the third call instruction is encountered, so as the end of the three instructions the stack pointer has been decremented to address 1FFAH and the stack is as shown in *Figure 5.8*.

The three return addresses appear on the stack in descending memory addresses. In other words the stack appears to be 'hanging' in memory rather than building up as in the plate analogy. However, this makes no difference to the operation of the stack. The essential point is that the

stack pointer is pointing to a last number placed on the stack.

When the **return** instruction is encountered at the end of each subroutine, the stack pointer is pointing at the last address stored. This address is taken from the stack so that the program can continue with the instruction after the last **call**.

When the RET instruction is encountered, the following mechanism takes place:

(a) The data from the memory address pointed to by the stack pointer is read into the Z register in the CPU.
(b) The stack pointer is incremented.

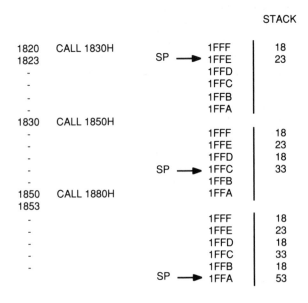

Figure 5.8   'Nested subroutine' stack operation

(c) The data from the address pointed to by the stack pointer is read into the W register in the CPU.

(d) The stack pointer is incremented again.

The result is that the stack pointer is incremented twice, so it is already pointing at the next unused **return** address on the stack. Also the first return address is in the W and Z register pair and this is then used instead of the program counter for the next machine cycle. This then forces the program to return to the instruction after the last call.

5.4 SAVING DATA ON THE STACK

The stack is clearly a very useful mechanism within the microcomputer system to store addresses temporarily, while subroutines are being executed. However, this is by no means the only use to which it may be put, since it is also possible to store data from registers in the CPU on the stack whenever required. The mechanism is almost identical except that it is initiated by the programmer using a **push** instruction. Data may be retrieved with a **pop** instruction.

When subroutines are used in a program, they often perform complex functions which require a lot of processing power. Generally this means that they require the use of most of the CPU registers. If these registers were in use in the main program which **called** the subroutine, there would obviously be a conflict of interests, since the data in the registers before the subroutine was called may need to be used after the subroutine. If it were destroyed, the whole function of the program could be lost. In circumstances likes this, the obvious answer is to save the data in the registers before a subroutine is called, and then restore the original data when the subroutine is complete. There are also occasions when the data in the registers at the beginning of a subroutine may be used by the routine in various calculations. This is a process known as **parameter passing** and is discussed further later in this chapter.

The instructions used to save and restore the contents of registers are known as the **push** and **pop** instructions respectively in the Z80. In other microprocessors they may have different names such as **push** and **pull**, but their function is very

similar. In the Z80 the **push** and **pop** instructions operate only on register pairs, and for that reason they are to be found in the 16-bit **load** group of instructions shown in *Figure 5.9* (page 104).

Note that the push and pop instructions use register indirect addressing. This is because the stack pointer acts as the register that holds the memory address into which the data is put. The correct mnemonics for the instructions utilise the register pair name as would be expected, and some examples are given below:

PUSH AF	PUSH accumulator and flags onto the stack.
PUSH DE	PUSH the D and E registers onto the stack.
POP HL	POP the top of the stack into the H and L registers.

This syntax for all the other instructions is similar to that shown above. Note that the stack pointer is automatically decremented after each **push** instruction by 2, and it is automatically incremented by 2 after each **pop** instruction. The push instructions take the register contents and place them on the stack with the high byte first followed by the low byte, while the pop instructions pop the low byte first followed by the high byte. With the pop instruction, there is no record of where the data came from which is found on the stack. The data is simply transferred from the top of the stack to the data registers chosen in the pop instruction.

Operation of the Stack Pointer with Push and Pop Instructions

The stack pointer works in more or less the same way with the **push** and **pop** instructions as it does with the operation of a subroutine **call**. This is shown below.

When a push instruction is executed, for example push HL, the sequence of operations is:

(a) The stack pointer is decremented.

(b) The contents of H are written into the address pointed to by the stack pointer.

(c) The stack pointer is decremented again.

(d) The contents of L are written into the address pointed to by the stack pointer.

SOURCE

DESTINATION		REGISTER							IMM. EXT.	EXT. ADDR.	REG INDIR.
		AF	BC	DE	HL	SP	IX	IY	nn	(nn)	(SP)
REGISTER	AF										F1
	BC								01 n n	ED 48 n n	C1
	DE								11 n n	ED 58 n n	D1
	HL								21 n n	2A n n	E1
	SP				F9		DD F9	FD F9	31 n n	ED 78 n n	
	IX								DD 21 n n	DD 2A n n	DD E1
	IY								FD 21 n n	FD 2A n n	FD E1
EXT. ADDR.	(nn)		ED 43 n n	ED 53 n n	22 n n	ED 73 n n	DD 22 n n	FD 22 n n			
REG. IND.	(SP)	F5	C5	D5	E5		DD E5	FD E5			

PUSH INSTRUCTIONS →

NOTE: The Push and Pop Instructions adjust the SP after every execution.

POP INSTRUCTIONS ↑

Figure 5.9 PUSH and POP instructions

Similarly, when a pop instruction is executed, such as pop BC, the sequence of events is:

(a) The contents of the address pointed to by the stack pointer is written into register C.
(b) The stack pointer is incremented.
(c) The contents of the address pointed to by the stack pointer is written into register B.
(d) The stack pointer is incremented.

It is important to realise that the same stack is being used whether it is the contents of a register pair or a subroutine return address which is placed on it. In fact the data on the stack will usually be a combination of both, mixed depending upon the sequence of operations in a program. Therefore it is important that the programmer realises the order of data on the stack, so that when information is 'popped' from it, only the right number of bytes are removed, so that the subroutine return addresses will not be lost. In general therefore, every push instruction must have a corresponding pop.

The order in which data is popped from the stack is also important, since the programmer must remember that it is always the last data on which must be the first data off. When a number of register pairs have to be pushed onto the stack therefore they must be popped in the *reverse order*. The instructions in the subroutine would therefore be:

```
PUSH BC
PUSH DE
PUSH HL
  –
  –
  –
POP HL
POP DE
POP BC
```

The position of the push and pop instructions in a program is also significant. Consider the examples shown in *Figure 5.10*.

In Example (a), the subroutine uses the B, C, D and E registers in its calculation, and therefore the contents of BC and DE are saved on the stack prior to the subroutine **call**. After its return the contents of BC and DE are restored in reverse

order before the rest of the program proceeds. Example B uses the same subroutine, but this time the push and pop instructions are placed at the beginning and end of the subroutine respectively.

At first sight it may seem that there is very little difference between the two alternative methods of saving the register contents that are to be used in the subroutine. However, on closer examination, it is clear that method B is a far better programming technique. This is because the register contents are saved in the subroutine, which means that the main program can be reduced in length and complexity. Since the subroutine is likely to be **called** a number of times the push and pop instructions are required only once if they are placed in the subroutine, whereas they would be required every time the subroutine was called if they were used in the main program. This would be very wasteful of space and the program would be much longer and less readable. It is also good programming practice to write subroutines so that they destroy no register contents if this is at all possible. This will allow greater flexibility in the writing of routines since there will be no need to remember the special features of each one, and which registers they destroy or preserve.

While it may be good practice that all subroutines do not change the contents of any registers on exit, it is not always possible to accomplish, especially when the subroutines have been written by other programmers. When specifying a subroutine it is vital that its effect is noted and any changes that occur in registers other than those that are specified, so that if the routine is used by another programmer all the information is available. One of the major benefits of utilising subroutines is that it is possible to use those written by other people to save a lot of work. This is the case where a programmer uses routines that are already written as part of a system firmware, such as the routines found in the MONITOR program.

```
MAIN    -                        MAIN    -
        -                                -
        -                                -
        PUSH BC                          CALL SUB
        PUSH DE                          -
        CALL SUB                         -
        POP DE                           -
        POP BC                           -
        -                                -
        -                                -
        -

SUB     -                        SUB     PUSH BC
        -                                PUSH DE
        -                                -
        -                                -
        -                                -
        -                                -
        RET                              POP DE
                                         POP BC
                                         RET

        (a)                              (b)
```

Figure 5.10 PUSH and POP instructions in subroutines

5.5 PARAMETER PASSING

The use of subroutines can considerably enhance the quality of computer programs, while at the same time making it possible to devise libraries of commonly used routines. The widespread use of subroutines makes it easier for the programmer to think in terms of the overall program concept rather than being bothered with the details of every instruction from an early stage. The logic of the program can be sub-divided into suitable subroutines which may be written later, or even by members of a programming team.

So far most of the subroutines considered have been fixed in their operation. They simply create a delay, or perform the same function every time they are used. They can be made far more useful if the precise operation of the subroutine is somehow controlled from within the main program. For example, a subroutine that multiplies two fixed numbers together would be of little value if the numbers were always the same whenever the routine was called. A multiply routine has to be able to accept different values sent from the main program and then return to the main program with the result of the calculation. Similarly, while a fixed delay program has some benefits, often it is desirable to have a delay routine that depends upon a number calculated from within the main program.

These two examples illustrate the need to be able to pass values or 'parameters' to subroutines so that they can act upon data that is present in the main program. There are a number of ways in which this operation may be performed, depending upon the circumstances. The most common are:

(a) Using the CPU registers.
(b) Using the stack.
(c) Using selected memory addresses.

Parameter Passing in Registers

Using registers to pass parameters is probably the simplest of all the methods, and also the most common. However, it does restrict a number of values that may be passed to a subroutine to those that can be contained within the register block and is only therefore suitable for a small number of parameters. In addition, care must be taken that the registers to be used are well documented, so that if the subroutine is used by another programmer, it will be obvious where the parameters may be found.

Consider the following example which is a variable delay routine, in which the length of the delay is a value passed in the D register:

```
MAIN:        LD D,30H
             CALL VARDEL
             _
             _
             _
             _
             LD D,80H
             CALL VARDEL
             _
             _

VARDEL:      LD BC,0000H
VARDE2:      DEC BC
             LD A,B
             OR C
             JP NZ,VARDE2
             DEC D
             JP NZ,VARDEL
             RET
```

The VARDEL program is called twice from the main program, each time with a different value loaded into D register just before the **call** instruction. In the subroutine, the main delay is simply a BC register pair being decremented to 0 but each time this occurs, before the subroutine returns the contents of the D register are also decremented so that it is only when this register is 0 that the subroutine returns to the main program. In this way the contents of D register affect the delay caused by the subroutine, and the value in the D register is said to be passed to the subroutine.

Values may also be passed from the subroutine back to the main program, such as when a subroutine is used for input of information. For example, an **analog** subroutine may be used to obtain the value of an analogue input and pass it back to the main program to be used on the display and for calculation purposes. Typically this single value could be passed in a register.

Passing Parameters on the Stack

Another parameter passing technique, used frequently in the implementation of high-level languages, is to pass the values on the stack. To do this, the values held in registers are simply pushed onto the stack before the subroutine is called, and any resulting values are popped from the stack at the end of the subroutine. While this may seem a simple enough method, it has a number of operational difficulties in a microcomputer system, because the stack is also used for the return address for the subroutine, and care must be taken not to mix the parameters with this address.

One way of achieving this is to directly manipulate the value in the stack pointer at the beginning of the subroutine, so that the return address is avoided. This involves decrementing the stack pointer twice so that it points to the parameters and misses the return address. However, care must be taken to return the same number of parameters to the stack that are taken from it, and also to increment the stack pointer twice, to return it to the correct value so that it returns correctly from the subroutine.

This operation is illustrated below using part of a multiplication routine. The two numbers to be multiplied are stored in the B and C registers which are then pushed onto the stack before the multiply operation. On return the result is also stored on the stack, and this is popped back into the BC pair after the subroutine has returned to the main program. In the subroutine itself, the values previously stored on the stack may be popped into any convenient register pair, and here the HL pair is shown, for operation during a subroutine. There is no obligation to use the same registers that were used in the main program. As long as there are the same number of DEC SP instructions as there are INC SP instructions, and the same number of **pushes** as there are **pops**, then this method works very efficiently. The routine shown below is for a multiply operation, but may equally easily be used for any type of subroutine, although it is particularly good with mathematical operations:

```
MAIN:   LD B,n      ; FIRST NUMBER
        LD C,m      ; SECOND NUMBER
        PUSH BC     ; VALUES TO STACK
        CALL MULT
        POP BC
        –
        –
        –
```

```
MULT    INC SP
        INC SP
        POP HL      ; GET PARAMETERS
        –
        –
        –
        –
        –
        PUSH HL     ; RESULT TO STACK
        DEC SP
        DEC SP
        RET
```

A variation on this theme of using the stack to store the subroutine parameters, is to use alternate stacks for subroutine return addresses, and the values to be stored. Once again this technique is not difficult in practice, it simply requires the correct value of the stack pointer to be loaded before parameters are **pushed**, and then before the subroutine is **called**, and the same thing to take place at the beginning and end of the subroutine.

Once again the use of the stack in passing parameters to subroutines tends to be restricted to relatively small numbers of parameters, although in theory it is possible to use the technique for a very large number of parameters. There would be a speed penalty in this, however, because of the large number of **push** and **pop** operations which would be required.

Passing Parameters in Memory

For large numbers of parameters to be used in subroutines, the only feasible method is to use a dedicated area of memory to store them. Often, instead of passing all the values to the subroutine, a register pair is used to simply pass the value of the start address of the memory block where they are stored. In addition, if the length of the block is not fixed, another parameter holding the number of bytes may also be passed to the subroutine. This technique is particularly useful when, for example, messages in ASCII code need to be passed to a subroutine for output, or large numbers of values need to be manipulated mathematically. Consider the case of where 200 numbers must be output to an output port in rapid succession, such as when an analogue waveform must be generated from digital data. If the start address of the numbers in memory is held in a register pair, and the number of bytes to be output in another register or register pair, then the program that could be used to perform this operation would be similar to that shown below:

```
MAIN:   LD HL,1900H  ;  START OF DATA
                        BLOCK
        LD BC,00C8H  ;  200 BYTES
        CALL SEND
        –
        –
        –
        –


SEND:   LD A,(HL)    ;  GET BYTE
        OUT (81H),A  ;  OUTPUT IT
        INC HL
        DEC BC
        LD A,B
        OR C
        JP NZ,SEND
        RET
```

Notice that the value in the HL register pair is the start of the data block, and this is immediately used in the subroutine to get the first byte of data and output it. Within the subroutine the pointer is incremented so that it is able to progress through the whole data block, and the counter value in the BC register pair is decremented until the whole block is sent.

When large blocks of data are to be manipulated in this way in a subroutine, passing the parameters relating to the data block location in memory and its length, are the only realistic ways to proceed.

Summary

The most important points covered in this chapter are:

- **A subroutine** is a very important feature of good computer software. They allow programs to be broken down into smaller sections which make the writing and testing of them much easier.

- Subroutines allow smaller programs to be written since they only need to be entered into memory once even though they are used many times. Their only disadvantage is that there is a small penalty associated with the additional instructions needed to **call** and **return** from them.

- The microprocessor has an automatic mechanism which stores the correct return address each time a subroutine is **called**. This involves the use of the **stack pointer** register and an area of memory known as the **stack**.

- The stack is a **last-in, first-out** storage device, in which the last data stored is always the first data to be retrieved.

- The stack can be used to store the data from registers in the CPU and this is controlled with the **push** and **pop** instructions. Care must be taken by the programmer however, so that data and subroutine return addresses are not mixed up by the processor.

- Subroutines can be made more versatile if parameters are passed from the main program. In this way the basic function of a routine can be adjusted, or it can be made to manipulate different data.

Questions

5.1 Write down the hex code for the instructions given in mnemonic form below:

(a) CALL NZ,19B3H.

(b) CALL 5432H.

(c) RET Z.

5.2 If the stack pointer in a microcomputer system is loaded with the value 4010H, what will its contents be if two **call** instructions are executed one after the other?

5.3 Briefly explain how the return address for a subroutine is automatically available in the microprocessor during the execution of a call instruction.

5.4 What is likely to limit the size of the stack in a practical microcomputer system?

5.5 Write down the hex code for the following instructions:

(a) PUSH IX.

(b) POP AF.

(c) PUSH BC.

5.6 Write down the effect of the following instructions if they occurred as part of a program.

PUSH AF
POP BC

5.7 What is the effect of the following sequence of instructions?

PUSH AF
POP AF
PUSH AF
POP AF

5.8 Briefly explain how it is possible for a microcomputer system to tell the difference between the contents of a register pair and the return address for a subroutine which are both found on the stack.

5.9 Briefly describe the purpose of using parameter passing with subroutines.

5.10 What are the advantages of using the stack to pass parameters to the subroutine?

5.11 List at least three types of subroutine which may benefit from use of parameter passing.

5.12 What are the main advantages of the use of subroutines in software?

5.13 What is meant by a 'nested subroutine'?

5.14 What is a LIFO stack?

5.15 If the stack pointer value is 1F05H in a system, what would it be if the instruction POP BC were executed?

Answers

CHAPTER 1

1.1 (a) The Telephone System.
  Inputs – Electric power, soundwaves.
  Outputs – Soundwaves.
  Process – Small-amplitude soundwaves are transmitted electrically over long distances and re-converted to soundwaves at a distant receiver.
  (b) A radio receiver.
  Inputs – Electric power, electromagnetic signals.
  Outputs – Soundwaves.
  Process – Small-amplitude electromagnetic waves, suitably modulated, are converted into sound energy.
  (c) A domestic toaster.
  Inputs – Electric power, bread.
  Outputs – Toasted bread, waste heat.
  Process – Bread is heated until it is turned brown, at which point it is ejected. Waste heat is generated as a by-product.

1.2 (a) Analogue.
  (b) Digital.
  (c) Digital.
  (d) Analogue.

1.3 Normally, zero volts represents a logic 0 and +5 volts represents a logic 1. However, there is some latitude in these voltages and in practice a logic 0 may be between 0 and 0.8 volts while a logic 1 may be between 2.4 and 5 volts.

1.4 An analogue-to-digital converter is a device that changes an analogue signal into a digital signal which may be used as the input to a computer. It allows 'real world' signals to be processed in the world of digital electronics.

1.5 Binary numbers may be used to represent:
  (a) numbers,
  (b) alphanumeric characters (ASCII Code),
  (c) computer instructions.

1.6 (a) C5,
  (b) F1,

(c) 0 0 0 1 1 1 0 0
(d) 1 0 1 0 1 1 1 1

1.7 Serial data transmission takes places one bit at a time over a single wire. This is a particularly slow method of data transfer. Parallel data transmission requires one wire for each bit of data, e.g. 8 wires for an 8-bit byte. However, it is much faster than serial data transfer since all the bits are transferred simultaneously.

1.8 ASCII code is used to represent alphanumeric characters, punctuation marks and control instructions for computers and peripheral devices. This universal code allows text to be manipulated easily in computer systems.

1.9 The address bus is derived from the microprocessor and is used to hold the binary number which is the unique address of the part of the system which is required to operate at any moment.

1.10 The data bus is described as bidirectional since it carries data both to and from the microprocessor.

1.11 Some instructions take longer than others to execute because they require extra information to be retrieved from the memory before they can be completed. Simple instructions are completed within the microprocessor itself while others need to read memory or input or output data.

1.12 The main function of a register is to store binary numbers or data. Generally, a register stores one byte (8 bits).

1.13 A register requires two control signals so that it can be loaded and read independently. One control signal loads the register and another control signal reads the data from it.

1.14 A multiplexer is used to connect the microprocessor registers to the data bus to simplify the connection problem. Generally it is only necessary to have the ability to connect a single register to the bus at any time, so a multiplexer provides an ideal method of connection.

1.15 The flags in a CPU change state as a result of any arithmetic or logical operation that takes place within the arithmetic logic unit.

1.16 Calculations performed by the ALU are generally placed back in the accumulator. These calculations also affect the flag register.

1.17 A microprocessor steps through its program automatically because of the action of the program counter. This stores the start address of the program and then each time it is used its value is automatically incremented so that it is ready to receive the next byte of data or instruction from the next memory location in sequence.

1.18 The general purpose registers within a microprocessor are used to store intermediate calculations or values which are due to be processed by the CPU.

1.19 A system with 24 address lines could address directly 16 777 216 addresses.

1.20 The best type of ROM to use would be an EPROM.

1.21 Static RAM requires no refresh circuitry to be included in the system and in addition certain types of static RAM are faster in operation than dynamic RAM.

1.22 The program counter in a Z80 is a 16-bit register which keeps track of the current memory address from which data or instructions are to be read. Each time the program counter is used its value is automatically incremented so that it is ready to obtain data from the next address during the next machine cycle.

1.23 (a) 18F1
    (b) 85AC
    (c) 7F36.

1.24 (a) 0 0 0 1 1 0 0 1 1 0 0 0 1 0 0 1
    (b) 0 1 0 1 0 1 1 0 1 0 1 0 1 1 0 1
    (c) 1 1 1 1 1 0 1 0 1 0 1 1 0 1 0 0

1.25 Read only memory is necessary in a microprocessor system since it is the only type of memory that stores its data even though power is removed. Therefore, the programs are ready to be operated as soon as a microcomputer is switched on. All microcomputers require some start-up programs and these are known as the monitor program and are stored in read only memory devices.

CHAPTER 2

2.1 The control bus is responsible for telling the rest of the computer system HOW to operate during any machine cycle. For example, it indicates whether memory or input/output devices should operate, and whether the cycle is to read or to write data.

2.2 An op-code is the first part of an instruction. It contains the coded information that indicates to the microprocessor which instruction is required.

2.3 The **clock** in a microprocessor system is usually a stable squarewave. It is used to synchronise the operation of the complete microprocessor system.

2.4 (a) A T-state is the name given to one cycle of the microprocessor clock.
    (b) A machine cycle is the name given to one cycle of the internal microprogram which runs the microprocessor.

2.5 The number of T-states and the machine cycles an instruction will require for completion is determined by the nature of the instruction itself. Some instructions require only one machine cycle since their action is purely internal whereas others require extra machine cycles to obtain data from the system memory or input/output ports.

2.6 Built into each instruction is a code that tells the microprocessor how many bytes are required and what function each will perform. This is part of the op-code.

2.7 The instruction LD (1950H),A would require 4 machine cycles. These would be instruction fetch, followed by two memory read and one memory write cycle.

2.8 The $\overline{\text{IORQ}}$ signal and the $\overline{\text{WR}}$ signal would be active during an output write cycle in a Z80.

2.9 The time delay between the address and the control signals being applied to a memory device and the data becoming available is known as the 'memory access time'. This is the time it takes for the memory to respond with the data. Typically the time is between 50 and 450 ns.

CHAPTER 3

3.1 The instruction set of a microprocessor is the complete range of instructions that it can execute. This may vary considerably from one microprocessor to another depending upon its design.

3.2 The main classes of instruction found in all microprocessor instruction sets are:
    (a) Data transfer instructions.
    (b) Data manipulation instructions.
    (c) Flow of control instructions.
The flow of control instructions are sometimes known as the **test and branch** instructions.

3.3 Assembly language programming is often preferred to a high-level programming language because it can be made to operate at higher speed, and uses less memory.

3.4 A symbolic address is a label or name that is used instead of the hexadecimal value of a memory address. Often in assembly language programming the actual address to be used is not known when the program is being written so a symbol is used instead, consisting of a name or label for that address.

3.5 A **monitor** program generally performs the following functions:

(a) Interaction with keyboard and display.

(b) Machine code entry facilities into memory.

(c) Execution of machine code programs.

3.6 If the microcomputer system is to be used for assembly language program it must have:

(a) A full alphanumeric keyboard.

(b) A display for alphanumeric information.

(c) An **editor** program.

(d) An **assembler** program.

(e) A **monitor** program.

3.7

Mnemonic	*Hex code*
(a) LD C,H	4C
(b) LD A,L	7D
(c) LD L,A	6F

3.8
(a) 26 37
(b) 1E 01
(c) 2E 76

3.9
(a) LD A,59H
(b) LD E,1EH
(c) LD B,B1H

3.10

Mnemonic	*Hex code*
(a) LD D,(HL)	56
(b) LD C,(HL)	4E
(c) LD (HL),A	77
(d) LD (HL),67H	36 67

3.11 The sequence of instructions would transfer the data from address 1800 hex and put it in address 1900 hex.

3.12

Mnemonic	*Hex code*
H LD A,(1910H)	3A 10 19
LD (1920H),A	32 20 19

3.13
(a) LD A,(1905H)
    LD C,A
(b) LD HL,1905H
    LD C,(HL)

3.14 LD DE,2077H.

3.15

Mnemonic	*Hex code*
LD HL,1800H	21 00 18

3.16 The instruction loads address 1234H with the contents of the C register, then address 1235H with the contents of the B register. Its hex code is ED 43 34 12.

3.17

Mnemonic	*Hex code*
(a) OUT (63H),A	D3 63
(b) EX DE,HL	EB
(c) LD B,53H	06 53
(d) LD BC,6417H	01 17 64
(e) LD HL,(6417H)	2A 17 64
(f) LD (6417H),A	32 17 64

3.18
(a) 0 1 1 0 1 0 0 1
(b) 1 1 1 0 1 1 1 0
(c) 0 1 0 1 0 1 1 0

3.19
(a)

0 1 0 0 1 1 0 1	+77
0 1 1 0 1 0 0	+52
1 0 0 0 0 0 0 1	−127

Note that this results in a sign error because the result is greater than +127 if signed number representation is used.

(b)

0 0 0 1 1 0 0 1	+25
1 1 1 1 0 0 1 1	−13
0 0 0 0 1 1 0 0	+12

(c)

0 1 0 1 0 0 0 1	+81	
1 0 1 1 1 0 0 1	+70	INVERTED
1		
0 0 0 0 1 0 1 1	+11	

(d)

1 1 1 1 0 1 0 0	−12	
0 0 1 0 0 0 0 0	−33	INVERTED
1		
0 0 0 1 0 1 0 1	+21	

3.20

Mnemonic	*Hex code*
(a) ADD A,E	83
(b) ADC A,81H	CE 81
(c) ADD A,(HL)	86

3.21
(a) LD A,B
    ADD A,20H
    LD B,A
(b) IN A,(80H)
    ADD A,H
    LD H,A.

3.22

Mnemonic	*Hex code*
(a) SBC A,(HL)	9E
(b) SUB A	97

3.23 LD A,L
SUB 5H
LD L,A

3.24

Mnemonic	*Hex code*
(a) INC A	3C
(b) DEC (HL)	35

3.25

	Mnemonic	Hex code
(a)	SBC HL,DE	ED 52
(b)	DEC DE	1B
(c)	ADD IX,IX	DD 29
(d)	DEC SP	3B

3.26 (a) AND D

(b) OR (HL)

(c) XOR FEH

3.27 The result would be 0 1 1 1 0 0 1 0, in binary.

3.28 The XOR A instruction performs an Exclusive OR with the accumulator itself, which always results in a value of 0 0 0 0 0 0 0 0 in the accumulator. It also has the effect of clearing the carry flag.

3.29 IN A,(80H)

AND OFH

RLCA

RLCA

OUT (81H),A

HALT

3.30 The carry generated as a result of the addition of two bytes of data must be added in when the two next most significant bytes are added. This is achieved by employing the ADC instruction.

3.31 If the decimal adjust accumulator instruction is included in the program after each binary mathematical operation, the numbers will be adjusted by the processor to standard binary coded decimal format.

3.32 (a) AND E3H

(b) OR C2H

(c) XOR 28H

3.33 F6 Hex.

3.34 Bit masking is the process whereby certain bits of a register are forced to a logic 0 or logic 1 state so that they will be in a known condition.

CHAPTER 4

4.1 An algorithm is a sequence of steps that together describe the method to be used to solve a problem.

4.2 It is better for a computer to calculate the values to be output to produce an analogue waveform if they have a simple mathematical relationship with each other.

4.3 Two essential features of a MONITOR program include:

(a) The ability to allow a machine code to be entered into the computer memory.

(b) The ability to execute a machine code program from a given address.

4.4 A flow chart is a diagrammatic representation of the steps in an algorithm. These steps show the flow of instructions necessary to solve the problem for which the algorithm was written. A flow chart is generally an intermediate step between the definition of the algorithm for a computer program, and the instructions that result from it.

4.5

PC	A	C	D	E	H	L	1B00	1B20
1900	00	00	00	00	00	00	00	00
1903	00	00	00	00	1A	03	00	00
1905	00	00	00	00	1A	03	77	00
1907	00	00	18	00	1A	03	77	00
1909	00	00	18	03	1A	03	77	00
190A	00	77	18	03	1A	03	77	00
190B	77	77	18	03	1A	03	77	00
190C	77	77	18	03	1A	03	77	77
190D	77	77	18	03	1A	03	77	77

4.6 A BREAKPOINT is used in software development to stop the execution of a program when it reaches a specific address. This allows the programmer to examine registers and/or memory addresses to check the operation of the program.

CHAPTER 5

5.1 (a) C2 B3 19

(b) CD 32 54

(c) C8

5.2 400C hex.

5.3 The subroutine **return** address is the value held in the program counter after the complete **call** instruction has been read from memory. This is because the program counter is incremented during T2 of the third machine cycle of the instruction, and it is therefore automatically available to be put onto the stack by the CPU.

5.4 The theoretical size of the stack is limited by the amount of memory available in the system. In practical terms, the programmer may set aside about 100 bytes for stack space.

5.5 (a) DD E5

(b) F1

(c) C5

5.6 The contents of AF would be transferred to BC.

5.7 They cause a short delay.

5.8 The system has no way of distinguishing where data on the stack came from. This is the programmer's responsibility.

5.9 Parameter passing allows a subroutine to be used with different data while keeping its basic function fixed.

5.10 No registers are required.

5.11 Arithmetic routines, variable delays and variable input/output routines.

5.12 Subroutines have the following advantages:

(a) Programs are better structured.

(b) Memory requirements are reduced.
(c) Routines can be written and tested independently.
(d) Program libraries of subroutines can be written.

5.13 One subroutine **called** from another.

5.14 A last-in first-out storage device. The last data stored is the first retrieved.

5.15 1F07H.